THE FUTURE of BANANAS

'This is a fascinating exploration about not just the humble banana, but a fruit whose history encompasses colonialism, war, globalisation, drug trafficking, changing consumer tastes and scientific breakthroughs. Dale also makes an eloquent case for the importance of embracing genetically modified technologies – if we want to carry on eating the world's favourite fruit.'

– **HARRY WALLOP**, award-winning consumer journalist and broadcaster, and author of ***Consumed: How We Buy Class in Modern Britain***

'*The Future of Bananas* is a gripping, insider account of the race to save the world's favorite fruit from all but certain extinction. If you've ever eaten a banana, you should read this book.'

– **GEORGE ZAIDAN**, author of ***Ingredients: The Strange Chemistry of Plants, Poisons and Processed Foods***

'*The Future of Bananas* makes a compelling case for optimism despite warnings of an impending "bananapocalypse." Dale brings decades of scientific expertise to the banana's complicated history of exploitation and disease, showing how the industry has adapted before and can again.'

– **JOHN T. LANG**, Associate Professor of Sociology at Occidental College and author of ***What's So Controversial About Genetically Modified Food?***

'Written in engaging prose by a leading biotechnologist, *The Future of Bananas* provides an accessible entry into the complex world of banana biodiversity, genetics, and the challenges of overcoming plant disease.'

– **JOHN SOLURI**, author of ***Banana Cultures: Agriculture, Consumption, and Environmental Change in Honduras and the United States***

THE FUTURE of BANANAS

James Dale

MELVILLE HOUSE UK
LONDON

MIX
Paper | Supporting responsible forestry
FSC® C104608

THE FUTURE of BANANAS

First published in 2026 by
Melville House UK
Suite 2000
16/18 Woodford Road
London E7 0HA
and
Melville House Publishing
46 John Street
Brooklyn, NY 112001

mhpbooks.com @melvillehouse

A CIP catalogue record for this book is available from the British Library

UK ISBN 978-1-911545-81-1
US ISBN 978-1-68589-237-1

1 3 5 7 9 10 8 6 4 2

Printed in Denmark by Nørhaven, Viborg
Typesetting by Roland Codd

The authorized representative in the EU for product safety and compliance is Easy Access System Europe, Mustamäe tee 50, 10621 Tallinn, Estonia
gpsr.requests@easproject.com

Foreword

If, like me, you live in a developed country, chances are when you go to your local supermarket, the fruit and vegetable section will be prominent, usually directly as you enter. It gives us a 'virtuous' feeling of buying healthy and fresh. Taking centre stage in that section will be a large display of beautiful, pristine yellow fruit: bananas, the world's favourite fruit, available all year round and at a very reasonable price. Look around. There are maybe five different varieties of apples, four different peaches, three different cauliflowers – white, orange and purple – plenty of choice for potatoes but just one banana: Cavendish, the variety named after the Cavendish family, the Dukes of Devonshire. Of course, your bananas didn't come from the family's ancestral holdings in northern England. They are the fruits of tropical plants.

Some of you will have seen banana plants growing in botanic gardens and maybe private gardens. Those of you who have ventured to the tropics will have seen them growing everywhere; bananas are ubiquitous in such environments. And you may have sampled some of the local varieties and wondered why these fabulous different bananas aren't sold in your supermarket back home. The reason? Because export bananas are a commodity crop centred around a single variety – the Cavendish – worth in excess of $15 billion each year. This is an enormous business, particularly when compared with the little local markets in Asia and Africa. Grown in huge monocultures in Latin America and the Philippines, Cavendish growing conditions are finely tuned, with harvesting and transport supply chains designed to deliver to your supermarket a perfectly ripe, yellow banana, ready to eat that day.

There are literally hundreds of different banana varieties but only Cavendish has the set of characteristics necessary to be a major export fruit, yielding big bunches of pristine fruit year round which are robust enough to be transported by sea to different continents. There was a very similar export banana variety that preceded Cavendish called Gros Michel (nicknamed Big Mike); it all but disappeared by the

1950s, wiped out by Panama Disease. This was one of the great plant disease pandemics, and one I will talk about in coming chapters.

There is so much more to bananas than the yellow fruit that we are conditioned to recognise. Bananas were one of the earliest crops, farmed in Asia, more than six thousand years ago, eventually expanding out of Asia through Africa and finally the Americas. They have a colourful history of exploitation, corruption and political interference as entrepreneurs established export markets in North America and Europe. There is an amazing diversity of varieties: fruits of different colours, textures, tastes and aromas with a multitude of uses. Tiny bananas with paper-thin skins are super sweet; big starchy bananas are steamed for hours and mashed as the staple foods of East Africa; the orange-fleshed bananas of Micronesia are stuffed full of beta-carotene; the plantains of Africa and the Pacific are huge and delicious; Kayinja, a popular banana in Uganda, is used to make beer. There is even a relative of the banana called Manila hemp, grown in the Philippines for its long strong fibres for rope-making.

All of these bananas have one thing in common: they are all ancient varieties selected by farmers over many centuries from naturally occurring hybrids. They were never bred, or rather they were bred by Nature. They were almost certainly selected by humans because the fruits have no seed; they are sterile. Great for eating, but not so great for evolving. And now the world has moved on, but the banana hasn't. The environment where that original Cavendish grew is very different from today's. There was no monoculture, no global warming, no disease pandemics. And herein lies the challenge: the banana as we know it faces commercial extinction, due in large part to Panama Disease tropical race 4 (TR4) which is decimating plants globally.

Can science help to save our favourite fruit? Despite Cavendish having no seed, we now have technologies that can breed new banana varieties by the centuries-old but modified technique of crossing one banana with another. We even have technologies that can subtly alter the genetic makeup of bananas to make them resistant to diseases, or to add nutritional value – and this is where my expertise and interests lie.

I have been involved with banana research for more than forty years, from around the time that the advanced genetic technologies, particularly genetic modification (GM), were first being developed. In those days, I worked on viruses of plants and I was given the task of identifying the cause of a disease called banana bunchy top which nearly destroyed the fledgling Australian banana industry in the 1920s. It was always thought to be caused by a virus and, with my first postdoctoral fellow, Rob Harding, we demonstrated that it was true. Virtually all banana varieties were susceptible to this virus, so developing a resistant variety was going to be impossible by conventional breeding; there weren't any resistant parents! We made the fateful decision to see if we could develop virus-resistant Cavendish bananas by genetic modification.

This was the early 1990s and a time when no one had genetically modified bananas. However, in the second half of that decade, there was growing realisation that something had to be done for the future of bananas. I was in the fortunate position to be in a country with a small but very visible banana industry, and the Australian Banana Growers Council enthusiastically supported our early research. Of course, this was before the rapid rise in sentiment opposing GM that bubbled up in the late 1990s. After that, most of our research was funded by international government, non-governmental, philanthropic and commercial groups and as a consequence, I led projects in Asia (Vietnam, Malaysia, Thailand and the Philippines) and Africa (Uganda, Kenya, Malawi and Rwanda), as well as Ecuador. In parallel, Rob Harding led projects in the South Pacific (Fiji, Papua New Guinea and Samoa). Much of our time in these countries was spent conducting extensive disease surveys, and this gave me the opportunity to experience first hand the amazing diversity of banana cultures. From the poorest subsistence farmers eking out food production from small plots growing a multitude of crops to massive, mechanised monocultures of Cavendish, I saw the different ecosystems and environments that existed, and I became committed to the idea that saving the banana was much more than saving Cavendish. And it shouldn't just be about disease resistance: it was clear to me that conventional breeding might provide a replacement Cavendish, but advanced genetic technologies had the potential to not only save the Cavendish but many, many other varieties of bananas.

Since then we have made serious progress. My team and I have developed a genetically modified Cavendish banana called QCAV-4. It is resistant to Panama Disease TR4 and has been released for commercial production in Australia, becoming the first genetically modified banana released in the world. With our close colleagues in Uganda, our 'golden bananas' await release in East Africa. More about all of this later.

I think, and hope, we can ensure the banana has a future, despite the red tape and regulation that often slows and stops progress. So let's explore together what could lie ahead for the world's favourite fruit.

The Ancient Fruits of the Forest

Let me take you back at least 100,000 years, well before modern humans, *Homo sapiens*, had made their way out of Africa and commenced their long expansion into Asia. At that time, the rainforests of the Asian tropics were pristine, with a huge diversity of plant species, each evolved to fill different ecological niches. From modern-day northern India across to southern China and down through South-east Asia – including Malaysia, Vietnam, Indonesia and New Guinea – to the tip of Australia, bananas were a part of these amazingly diverse forests – in particular, two species.[1] *Musa balbisiana* grew in the more challenging environments dominated by monsoons interspersed with extended dry periods, while *Musa acuminata* preferred the more luxurious forests with regular rainfall patterns. These 'wild' or indigenous bananas were quite different to the ones we eat today. The plants visually looked very much like the domesticated varieties we recognise. They grew tall and straight, they had very large leaves and they produced bunches of fruit. We know this because there are still remnant populations of these growing in the wild. There was however a major difference: their fruit were full of hard, pebble-sized seeds. There was very little flesh. These seeds were a result of sexual reproduction and each seed was a unique new combination of alleles* of genes, half coming from the male (pollen) and half from the female. So sometimes the two alleles came from the one parent (self-fertilisation), and sometimes each came from a different parent (cross-fertilisation).

These wild bananas had a second form of reproduction. Once a plant produces a mature bunch, that stem dies and, from the base of that stem, one or more new stems grow. These are referred to as suckers and

* We often refer to having 'different genes'. For instance, Person One has the 'gene' for blue eyes and Person Two has the 'gene' for brown eyes. In reality, there is only one gene for eye colour in humans, but different versions known as alleles determine the final eye colour. So, when we say an offspring has half its genes from its mother and half from its father, we are really referring to half of the alleles from each. However, for simplicity, I shall refer to the commonly used terminology of 'genes' representing alleles.

they are genetically identical to the mother plant from which they grew. Thus, these wild bananas had two very different routes to reproduce: seeds that provided genetic diversity and the opportunity to be spread far and wide by birds or other animals, and suckers that maintained occupancy of that hard-won small patch of land, at the same time maintaining the identity of the mother plant. These bananas were maximising their prospects of surviving and thriving in the wild.

The strategy has worked. One small example: about twenty-five years ago, I was part of a collecting expedition in dense rainforest in a national park in northern Vietnam. We were looking for evidence of virus infections in wild bananas in their natural habitat. These were wild seeded bananas scattered, usually singly, throughout the forest. However, we came to an area where a huge tree had fallen, probably in the previous year, producing a hole in the thick canopy. There, filling this new opportunity for land and light, were literally hundreds of banana seedlings, derived from seed produced by a mother plant probably growing very close to the tree before it had fallen. These seedlings were true opportunists, grabbing that window to the sky to establish their survival.

While it is likely that the early hunter-gather colonisers ate these seeded fruits, it is also possible that they ate the flowers. Banana flowers are still part of many Asian cuisines and can commonly be found in Asian food markets. It is also likely that these early colonisers would have used the large banana leaves to make shelters, again a practice that continues today. Thus, even without the seedless fruit we know, wild bananas were already very useful.

They were also easy to transport and re-establish using their suckers. We know that *Musa acuminata*, in particular, evolved into a recognisably different, geographically isolated sub-species in different countries: for instance, *banksii* evolved in New Guinea, *malaccensis* in Malaysia, *zebrina* in Indonesia, *errans* in the Philippines and *siamea* in northern Thailand.[2] There were others. With early colonisers moving these sub-species between regions as suckers, it provided the opportunity for new encounters and there was sex between the local bananas and the newly arrived bananas. From these new encounters and new combinations of genes came the next generation. While we are not certain of the mechanism, some of these new combinations produced bananas whose fruit had very few if any seeds. This wasn't a result of human intervention: rather, these new bananas were the result of 'Nature's breeding programme', the massive worldwide random recombination of genes within, and sometimes even between, species. This is the process that creates new individuals, the cornerstone of evolution from which the fittest of these new individuals would thrive and complement or replace those coming before.

The original progenitors, the wild bananas, were diploid. That is, their cells contained two copies of each of their chromosomes. Humans are also diploid. It is probable that the early, new low-seeded bananas were diploid and some of these are still cultivated today. However, the third generation of bananas, also a product of Nature's breeding programme, were triploid; their cells contained three copies of each of their chromosomes. These are close to sterile; they have very few if any seeds. They are known as parthenocarpic as they produce fruit without seeds. Even though human involvement in developing parthenocarpic bananas was purely accidental, it is not difficult to see why these naturally occurring hybrids were quickly selected by early farmers. The 'new' bananas were easy to grow, produced fruit throughout the year and were a wonderful source of starch and sugars. They became an important food crop throughout the centres of origin. There were probably thousands of different parthenocarpic combinations and there may be still more occurring to this day. We are aware of more than a thousand with a huge level of genetic diversity, resulting in bananas of different shapes, colours, sizes, taste, texture, climatic adaptation and many more characteristics, the products of Nature's breeding programme. They were never deliberately bred; they were selected by the early farmers.

The first evidence for cultivation of bananas comes from the Kuk Swamp in Papua New Guinea, dating to 6,800 years ago.[3] Unfortunately, there is no DNA evidence that would allow us to determine whether they were seeded or not. However, there is genetic proof that Cavendish, the world's most popular banana, is more than 1,000 years old and has genes from *malaccensis*, *zebrina* and *banksii* sub-species. These are true 'heritage varieties'; they are 'the ancient fruits of the forest'.

The progeny of these first seedless bananas had to be grown from suckers as there were no seeds. These suckers were genetically identical to their mother plant. The same process continues today. These new varieties were genetically 'fixed', meaning there has been no further evolution other than the occasional mutation, so the fruit of Cavendish almost certainly tastes exactly the same as that first seedless plant grown more than 1,000 years ago. There are minor Cavendish variants such as Dwarf Cavendish, Williams and Grand Nain but to the uninitiated, it would be impossible to tell the difference between them, other than perhaps by the height of the plant. This uniformity together with seedlessness is the reason bananas have become such a successful crop. Add to this, particularly for Cavendish in the developed world, year-round availability and an appealing taste and texture at a very reasonable price, and it is understandable why they have become the world's favourite fruit. But the same characteristics of uniformity and seedlessness have also become the domesticated banana's Achilles heel. Our favourite fruit has become genetically locked in time, unable to adapt to a changing world.

Well before the perils of sterility and uniformity started to become evident, humans began moving seedless hybrid suckers out of their centres of origin in Asia, establishing bananas as an important crop in virtually all regions of the tropical world. There is general agreement that this began more than five thousand years ago, with Arab trading routes extending through India and down the east coast of Africa.[4]

Bananas have become a very important food source in sub-Saharan Africa, and in modern-day Uganda, Rwanda and Burundi they have become the dominant staple. In many parts of Uganda, cooking bananas completely dominate the landscape; individual land holdings are usually less than two hectares in size but they coalesce into a sea of bananas.

South from Uganda in Tanzania, a diploid variety, Mchare, is one of the more popular cooking bananas.[5] Interestingly, this diploid is very closely related to one of the parents of Cavendish. In West Africa, plantains predominate as the cooked bananas of choice; plantains are often referred to separately from bananas but they are just one segment of the vast banana diversity, and there are many different types of plantains.

The genetic diversity of bananas in Africa is significantly less than in the Asian centres of origin but much greater than we see in supermarkets in the developed world. For instance, in Uganda, though dominated by the East African Highland cooking banana, plantains are also popular.[6] Most small farmers will also grow dessert bananas including Bogoya (Gros Michel) and Sukali Ndizi (sugar banana) but rarely Cavendish. Kayinja, a banana specifically grown for beer making, is also very popular and on many plantations you will find a covered, hollowed-out log where banana beer is produced by open fermentation with wild yeasts. These home brews are highly prized by their brewers, who are eager to share with visitors.

As with much of sub-Saharan Africa, bananas have become an integral part of the culture in the Pacific Islands. Over two thousand years ago, the Polynesians ventured out of East Asia and clearly brought banana suckers with them.[7] Coming directly from one of the centres of origin, the result of these intrepid travellers' efforts was that there is a huge diversity of varieties and landraces[†] grown in the Pacific Islands, and many different cooking types as well as dessert types. Perhaps the most spectacular are the fruits of the Fe'i bananas of Micronesia, many of which have bright orange flesh, a result of very high levels of carotenoids, the precursors of vitamin A. Of course, the early settlers didn't bring all the different varieties found in the island nations at one time. There must have been countless introductions, the vast majority of which have not been documented.

The last great geographic push by this amazing fruit was into South and Central America and the Caribbean. It was almost certainly the Portuguese

† In this book, I will use 'varieties' as the more common synonym for the scientific 'cultivar'. 'Landraces' are varieties or cultivars that have been locally selected by traditional methods.

who facilitated the move of bananas further west, probably to the Canary Islands as early as 1450 to 1500.[8] Over the next fifty years they were brought to the Caribbean and Central America, and by 1600 they were well established in South America. No one would have predicted that this region would become the dominant centre of a massive export industry based around a single type of banana, just one out of thousands.

By the 1850s, bananas were an integral part of the agriculture of the wet tropics and sub-tropics in Asia, Africa, the Americas and the Pacific Islands. These bananas, whether cooked as a starch source or eaten raw when ripe and sweet, were all consumed domestically, within a few kilometres of production.

This was about to change.

The Global Banana

By the 1800s, bananas had reached their natural geographical limits. Nearly all wet tropical and sub-tropical regions had embraced them as an important food crop. Most were grown by subsistence farmers or smallholders and commerce would have been mostly limited to local markets. This was because of one major limitation of the banana: ripening time. Once a bunch is harvested, the process is rapid; very quickly the fruit goes from hard green to soft yellow and then to brown and inedible within a few weeks. So the time spent in travel from plantation to consumer had to be short. Thus, for the western hemisphere of North America and Europe, bananas were an exotic fruit. There were plant collectors, garden enthusiasts and botanic gardens that had collections of bananas from the tropics, but most required heated glasshouses to keep the plants alive and thriving. These would have produced bunches and for a lucky few the opportunity to taste this exotic fruit of the tropics. Despite this, there were intriguing, unexpected encounters that are difficult to explain. For instance, Thomas Johnson, a merchant, herbalist and botanist, apparently displayed a bunch of bananas in his shop in London on 10 April 1633.[9] It was reported that the bunch came from Bermuda. How it got to London in a state to be displayed is amazing. Perhaps it was harvested very green and immature.

By the second half of the 1800s, the situation began to change. One of the key drivers of change was the identification of a banana that had both excellent consumer appeal and at the same time the characteristics of a potentially commercial product such as big bunches of robust, transportable fruit. One such banana emerged in the first half of the 1800s when a French naturalist, Nicolas Baudin, brought some plants from South-east Asia and planted them in a botanic garden in Martinique.[10] Local farmers obviously recognised that this banana was clearly superior both as a plant and as a fruit to anything they already had and so it soon found its way across the island.

In the mid-1830s, a French planter, Jean Pouyat, also noticed the same banana and its popularity. He subsequently established a planting of it in Jamaica where it was quickly adopted by many of the local growers. This banana finally became known as Gros Michel: it was a tall plant that bore big bunches of fruit with quite thick skin. When ripe it was a brilliant yellow, and the fruit had excellent taste and texture. The fruit were also tough. The entire bunch could be harvested, loaded onto a wagon and transported to market with very little damage. However, they also ripened quickly, still limiting their availability to local markets within a few days' travel.

The second key driver was the development of steamships which allowed much shorter transit times. The combination of a commercially viable fruit and a method of getting that fruit to markets, particularly in the US, kick-started the rapid development of a new industry. It attracted innovators, risk-takers and above all ruthless and often corrupt entrepreneurs.[11] While there were probably others, a sea captain called Lorenzo Dow Baker was one of the earliest to try his luck with exporting bananas from the Caribbean to the eastern states of the US and quickly became the largest exporter from that region. He then ventured into production. In 1885, Baker teamed up with Andrew Preston, a US produce buyer.[12] Preston turned out to be quite an innovator as he was the first to introduce 'refrigerated' bananas. Not quite the sophisticated containers of today. Rather, the cooling was a variation on the old ice box using huge blocks of ice.

Very quickly it was realised that, in addition to the Caribbean, first Central and then South America offered ideal growing conditions for bananas, particularly Gros Michel: hot and humid with high rainfall and fertile soil, coupled with compliant national governments looking to develop export industries. Initially, there were relatively few 'players' in the establishment of what would become a huge export industry and this heavily impacted the economic and political development of the industry.

Enter Minor Keith, an entrepreneur who was a railroad builder.[12] His first adventure into bananas was in Costa Rica where he struck a deal with the President: he would build a railroad for free, and in return receive a ninety-nine-year lease on the line, 800,000 acres of land adjacent to the line, plus control of the port of Limon. On those acres he planted bananas, initially to feed his workers, but he soon realised that there was an incredible opportunity selling the fruit. There were downsides. Clearing the land for the track and the bananas resulted in the deaths of many workers from disease, particularly yellow fever and malaria. However, the banana crop was very commercially successful and Minor became a significant producer of export bananas. He then joined forces with Preston and between them they established plantations in Panama, Guatemala, Honduras and Nicaragua in Central America, and Colombia and Ecuador in South America. In 1899, they called their company United Fruit, a name that would become synonymous with ruthless business practices and political interference.

Separately, Sam Zemurray,[13] known as Sam the Banana Man, was establishing himself as a banana buyer and trader in the US, initially buying cheap nearly ripe fruit at port and transporting them rapidly by rail to retail markets close by. He then expanded into buying bananas from Central America and by 1903 was buying and distributing fruit produced by United Fruit. In 1905, with partner Ashbell Hubbard, he bought the Cuyamel Fruit Company and in 1910 acquired 5,000 acres in Honduras to grow bananas for export. He then bought out Hubbard and expanded dramatically in Honduras.

Trouble was brewing when, in 1910, the Honduran president, Miguel Davila, planned to collect taxes through customs offices to repay the country's external debt. Zemurray was fearful this could ruin his export business, so through recruited mercenaries and others he organised a successful coup that saw the previously exiled Manuel Bonilla re-installed as president in 1912. Needless to say, Zemurray did very well out of this coup as Bonilla gave him very favourable tax and land concessions.

In the same decade, however, Cuyamel and United Fruit were in near constant conflict and in 1929, Zemurray finally sold Cuyamel to United Fruit for $31.5 million in United Fruit stock. But Zemurray was not done. Later, when the value of his United Fruit shares had declined by nearly 90 per cent, he pulled off another coup and took control of United Fruit as president.

United Fruit wielded huge political influence. By the 1920s, banana exports had expanded to a point where they were immensely valuable to the economies of many Central and South American countries. Often, this influence had disastrous outcomes for the local populations. One of the most shocking events occurred in Colombia in 1928.[14] Some 30,000 United Fruit workers went on strike over poor conditions including wages. In response, on 6 December, the military opened fire on the protesters, killing more than a thousand. Despite such occurrences, banana industry involvement in the politics of Latin America would continue, often with US intervention; in 1953, the CIA apparently assisted in orchestrating a coup in Guatemala to oust President Jacobo Arbenz who was threatening to redistribute vacant United Fruit land to the local peasants. And as recently as 2024, Chiquita (previously United Fruit) was ordered by a US court to pay $38.3 million to seventeen plaintiffs, relatives of victims of Autodefensas Unidos de Colombia (AUC) violence in Colombia between 1997 and 2004.[15] The AUC was a violent right-wing criminal group established in opposition to the Colombian communist resistance groups such as the Revolutionary Armed Forces of Colombia (FARC). Chiquita paid AUC $1.7 million, claiming the payment to be extortion or 'protection' money. The jury did not accept this and as a consequence the judge ordered the payout to the victims' relatives.

There were and still are other major international banana companies. Standard Fruit (now Dole) was established in 1924 and became a serious competitor to United Fruit. The company was also involved in some of the dubious political activities known collectively as the Banana Wars. The company Fyffes dates back to 1888 with their first shipment of bananas to London. They have been major buyers and marketers in the UK and Europe, but only bought their first plantation in 2015. The other major banana marketing company is Fresh Del Monte which produces bananas in Guatemala, Costa Rica and the Philippines but also purchases from a number of independent growers.

While the story of Gros Michel in the Americas from the late 1800s to the first half of the last century was very much about commercial exploitation and the development of a large-scale banana export industry, the story of Cavendish in the 1800s was much more 'romantic'. The story began, probably more than a thousand years ago, with a random hybridisation event in Asia, very probably southern China but suggested by some to be Vietnam. We shall never know for certain. Obviously, it was selected by the locals as a superior banana and became popular in its native territories. Somehow, suckers of Cavendish found their way to Mauritius, possibly via Chinese emigrants.

Enter the first of two 'botanists' who were instrumental in establishing Cavendish as a future mega-variety. Charles Telford (1788–1833), an Irishman born in Belfast, was an ardent amateur botanist living in Mauritius who collected as many specimens of bananas as he could, presumably from ships stopping by the island.[16] In 1829, he sent two suckers of a dwarf banana to a friend in Britain, Robert Barclay (1751–1830), indicating in the accompanying letter that he, Telford, had been collecting as many bananas as he could and that the two suckers were from the most valuable he had gathered. Barclay was another keen gardener and amateur botanist.

Unfortunately, Barclay died before he could see the fruits of his newly acquired specimens. After his death, the two plants were bought by Messrs Young of Young Nursery, Epsom. One of those plants was then bought by Joseph Paxton (1803–65), the head gardener at Chatsworth House in Derbyshire, for the princely sum of £10, or about £1,500 today, a transaction that the nursery later regretted. Chatsworth was owned by the 6th Duke of Devonshire, whose family name was Cavendish. Paxton built a glasshouse for the dwarf banana, where he multiplied the plants and took them through multiple fruit harvests. He and the bananas enjoyed considerable renown as a result, and in 1837 he published the first description of this dwarf banana, calling it *Musa cavendishii*. From these early beginnings, Cavendish commenced its lengthy journey to 'world banana domination'.[17] John Williams, a missionary, took Cavendish suckers to the South Pacific where they spawned a local banana industry although the unfortunate Williams was killed and cannibalised by the locals.[18] A third keen botanist, Alfred Diston, took three Cavendish suckers to Tenerife in the Canary Islands in 1846 and these were the foundation of a significant local industry. From there, Cavendish made its way to the Americas.

In 2019, I visited Faye Tuffrey, the then joint head gardener at Chatsworth House. I was privileged to have a tour of the Dwarf Cavendish plants held in a secure glasshouse in the grounds, not open to the public. (This isn't the original Paxton glasshouse, which was blown up early last century because it used too much coal for heating. Paxton also designed another grander glasshouse, Crystal Palace, which was to meet a similar fate, being destroyed by fire.) We are assuming that the plants in this secure glasshouse are direct descendants of that original sucker. I collected some leaf material from the plants and brought the material back to my lab where we extracted its DNA and sent it off for sequencing. When time permits, we will compare that sequence with other Cavendish DNA sequences from around the world. I'm not sure what we shall find. However, it is sure to be interesting.

Bananas remain an important point of interest at Chatsworth. On the second day of my visit, we went on a tour of the house. In one of the bedrooms, the guide specifically pointed out the banana image on the wallpaper. The story was that Paxton was inspired to grow bananas after seeing the image on this Chinese wallpaper. However, it wasn't until the 1920s that Duchess Evelyn moved a four-poster bed and found the image. The guide went on to say that Cavendish bananas were under threat from a disease, but that an Australian scientist was close to solving the problem. Much as I would like to think that the guide was one of the best-informed banana aficionados, I suspect that Faye might have alerted him to my visit. I did identify myself and we had an excellent discussion. On leaving the house, there were small Cavendish plants in pots for sale, although these were clearly not derived from the Dwarf Cavendish in the glasshouse. Bananas can be grown in Britain, but they usually struggle. I saw bananas in the wonderful Great Dixter garden in East Sussex developed by Christopher Lloyd. They were quite healthy, but you would need to be very optimistic to think they would produce big bunches.

Even before the burgeoning Gros Michel industry had reached its zenith, storm clouds were gathering. The first inklings of a pending disaster were not from the Americas but half a world away in Queensland, a British colony that would become part of the Federation of Australia in 1901, and which stretched from the tropics in the north to the sub-tropics further south around the capital, Brisbane. Queensland had a small, domestic banana industry. For a few years, there had been reports from local growers that their bananas were rotting and dying. Dr Joseph Bancroft (1836–94), a medical doctor and naturalist, was asked to investigate the possible disease on one plantation where Cavendish was thriving but a different variety, known as Lady finger, was dying. In his subsequent report, Bancroft described the symptoms of the disease we now know as Panama Disease, or more accurately Fusarium wilt of banana.[19] He believed the disease was caused by a fungus, as his microscopic examination revealed fungal threads (mycelium) in the diseased tissue. There were suggestions that the disease had been present around the Brisbane region for up to twenty-five years before Bancroft's investigation and report to the Queensland Legislative Assembly in 1876.

So why was it called Panama Disease rather than 'Brisbane Disease' or 'Queensland Disease'? In 1910, the American scientist R. E. McKenney published a paper in the prestigious journal *Science* titled 'The Central American banana blight', where he noted this serious disease in 1904 in Central American banana plantations, including some in Panama.[20] In the same year, another American scientist, Edwin Smith, published his own work in *Science*, demonstrating that a species of the fungus Fusarium caused an identical disease of bananas in Cuba.[21] Neither called the disease Panama Disease, nor did they recognise, or perhaps know about, the work of Bancroft. As they published their observations in a prestigious American journal, however, their works are usually recognised as the 'original' descriptions and the name Panama Disease came into common usage, as it still is today. I doubt whether the people of Brisbane or Queensland are too concerned about this historical oversight. The vast majority of plant diseases are named after their first identified host and their major symptoms. Hence, the most infamous of all plant disease epidemics – the one which caused the great Irish famine of the 1800s – is called late blight of potato.

In many ways, potatoes are similar to bananas. They evolved far away in the Andes of South

America; there are in excess of 4,000 different potato varieties and landraces, most of which are unknown outside the Andes; and they are grown from tubers not seed. And like bananas, they became a major food of Western countries. The Irish of the 1800s enthusiastically adopted potatoes as their primary, almost exclusive staple. There would have been extensive monocultures probably of only a few different varieties. Between 1845 and 1852, late blight, caused by the water mould *Phytophthora infestans*, hit the Irish crop causing the tubers to rot. There were massive crop losses, resulting in famine. During that period, around a million people died and a further million left Ireland.[22] Panama Disease was similarly to go on and wreak havoc on the burgeoning banana export industry.

To understand how a microscopic fungus and the disease it causes can produce such damage it is essential to go a little deeper. Panama Disease/Fusarium wilt of bananas is caused by the fungus *Fusarium oxysporum* f.sp. *cubense*; quite a mouthful, and so often referred to as simply Foc. This fungus lives in the soil and can survive for in excess of fifty years even without its favourite 'food', bananas. It survives probably as resting spores (seed) but may also colonise other plant species without causing any obvious symptoms. In the presence of bananas, Foc senses a meal and attaches to and enters the roots, whereby it grows up into the stem, killing the plant cells as it goes. It lives off this dead tissue. In the process, the fungus kills the plant's vascular system, starving the plant of water and nutrients. As a consequence, the plant wilts and usually dies.

It is the soil-borne lifestyle of Foc that has made it such a devastating pathogen. It almost certainly evolved along with bananas in Asia. Before human influence, bananas would have been spread by seed, probably via animals. However, once the seedless hybrids were selected as crop plants, it was usually the suckers that were moved, along with the soil attached to them: the perfect vehicle for a soil-borne pathogen. In the late 1800s and early 1900s, as Gros Michel cultivation was expanding at a great rate in the Americas, there was little real appreciation of the micro-organisms that inhabited soil and how they might be distributed on plants. We now know that there can be literally millions of micro-organism cells in as little as a gram of soil. So it is easy to understand how Foc hitched a ride on banana suckers around the tropical and sub-tropical world.

One other very important point: Foc exists as different types which we call races.[23] Panama Disease race 1 infects Gros Michel and was almost certainly the one encountered by Bancroft in Queensland that infected Lady finger. The other very important race is Panama Disease tropical race 4, or TR4.

As the vast monocultures of Gros Michel continued their expansion through Central America and into South America, along came Panama Disease race 1 for the journey. All the Gros Michel plants were genetically identical and, as a consequence, no resistant lines were found, but not for want of trying. All of the major producing countries in the Americas were affected by the early 1900s. To start with, only a few plants here and there showed signs; the fungus moves slowly. But by about 1925, it was obvious that the disease was a major and compounding problem. There were no chemicals that seemed to have any effect.[24]

The first strategy adopted by the large banana companies was to open up new land, clearing more virgin rainforest to establish new plantations. Of course, the fungus was in the soil attached to the suckers that were planted on this new ground. There was a delay in seeing symptoms in the new plantations but inevitably the disease made its presence felt. The next big idea was flooding, essentially turning the plantations into shallow dams for a period of time to 'drown' the fungus. This worked – for a while. Flooding certainly knocked back the populations of Foc but at the same time destroyed virtually all the other adapted micro-organisms living in that soil. But the disease reappeared, this time without much competition from other, beneficial organisms in the soil. The disease became rampant.

The writing was on the wall. By the 1950s, the industry in the Americas had run out of new land and the flooding strategy wasn't keeping the disease at bay. Dying plantations were abandoned and productivity in those remaining was in severe decline. The Gros Michel industry was dead.[25] And waiting in the wings was Cavendish. Even as far back as 1874, there were observations that Cavendish was resistant to Panama Disease.

Cavendish is very closely related to Gros Michel through their maternal line.[26] Like Gros Michel, it yields very well and the fruit has thick skin, facilitating easier handling and transport. It is not as tall as Gros Michel, which was an advantage for both harvesting and high wind resistance. Most importantly, from a western consumer perspective the Cavendish fruit was almost identical to that of

Gros Michel: it was about the same size and shape, when ripe it was beautifully yellow and the taste and texture was almost the same.

It was an 'off the shelf' replacement for Gros Michel. As quickly as Gros Michel had 'invaded' South and Central America, Cavendish came in and almost completely replaced it. However, the fruit was not as robust. It could not be transported 'on the bunch', as this resulted in too much damage. This led to one of the few major innovations in the banana industry: the banana box. The hands were and still are cut from the bunch in packing houses and packed into boxes of between 15kg and 18kg depending on the market. Refrigeration also underwent dramatic improvements to the point where the fruit was harvested green, the boxes loaded into refrigerated containers and the conditions manipulated for artificial ripening to ensure that when the fruit arrived at port, it would be fully yellow within a few days.

A much more recent 'innovation' has been to 'co-load' a more valuable product: cocaine. At a major banana conference a few years ago in South America, I gave the opening keynote presentation on Panama Disease and potential control strategies. I was followed by the local chief of police giving a speech on cocaine trafficking through the banana export trade.

He described the substitution of fruit boxes with boxes of cocaine, the containers with false walls, and the move to GPS tracking of pallets to ensure drivers didn't stop off on the way from plantation to port. It really was an eye-opener for me. Very recently, the Spanish police intercepted thirteen tonnes of cocaine as part of a banana shipment from South America.[27] Such intercepts are becoming more common.

So, for 150 years or more, the curved '6–8 inch long' bright yellow fruit has become synonymous with 'banana' in most developed countries. Even in the two largest producing countries, India and China, Cavendish is the dominant dessert banana. Cavendish has become the supermarket banana!

What about the rest of the world? During this rapid expansion of the Gros Michel industry and its subsequent demise in the Americas, there was very little commentary about the vast majority of banana production. Around 85 per cent of global banana production is grown and consumed domestically. While Gros Michel production was expanding in the Americas for export, plantains, a group of cooking bananas, were becoming increasingly popular with local populations in South and Central America and the Caribbean, as well as in Africa and Asia, where they were already an important food crop for workers and their families. Until the 1950s, there was very little information published about banana production outside the Americas – with the exception of Australia, which had both a domestic banana industry and a well-established agriculture research and development sector. After that period, agricultural scientists from developed countries began travelling more widely and many developing countries began to invest in establishing their own local agricultural services. As a result, Panama Disease race 1 was being reported in many countries and was infecting a range of different varieties. Gros Michel was out, Cavendish was in.

Crunch Time

After the first Panama Disease pandemic, there was a short calm period. It didn't last. There were two more big disease hits on the way.

In 1963, a previously unrecognised leaf disease of bananas was identified in the Sigatoka Valley in Fiji.[28] The disease was to become known as Black Sigatoka. It is caused by a fungus, *Pseudocercospora fijiensis*. This fungus invades the leaves and kills the tissue, with entire leaves turning dark brown to black before dying. With the loss of leaves comes reduced photosynthesis, leading to heavy loss of yield. It debilitates rather than destroys but if uncontrolled can reduce the harvest by more than 50 per cent. While it was first recognised in the Sigatoka Valley in 1963, it was probably already widespread throughout South-east Asia and the South Pacific. It moved rapidly, first into Honduras, Central America, in 1972 and then Zambia in Africa in 1973, and with footholds in those continents, quickly engulfed all the banana plantations in those tropical areas. The disease favours hot, humid conditions just as bananas do.

There are two major differences in the impact of Black Sigatoka compared with Panama Disease of Gros Michel. The first is that it affects a very broad range of banana varieties including Cavendish, Gros Michel, plantains, East African Highland bananas and many of the lesser-known varieties. It doesn't discriminate between the vast commercial monocultures of Cavendish in America and the smallholder subsistence farmers in Africa and Asia. The second difference does discriminate: the disease can be chemically controlled by regularly applying fungicides to the leaves. In high disease pressure areas, the fungicides need to be sprayed very frequently, often in excess of once a week.[29] Commercial exporters can and do afford these control measures; smallholder subsistence farmers cannot. Of course, there are consequences from these massive spraying regimes. There are well-documented impacts of the chemicals on the health of farm workers and their families, who often live around the plantations. The environmental impacts are also considerable, as the chemicals easily enter the waterways and surrounding areas. The Western consumer is usually oblivious to such issues: while growing, the bunches have plastic bunch covers placed over them so that the pristine yellow fruit arrives at market virtually unmarked and without residue.

The more recent organic banana sector avoids spraying with these chemicals by avoiding the disease as much as possible. Production is based in very dry climates not conducive to the disease. The crop must of course be irrigated rather than rainfed as in nature. This is expensive and also comes with an environmental cost.

The smallholder subsistence farmers often produce 'organic' bananas. They use manure as fertiliser and the rains for water, but without chemicals, Black Sigatoka slashes yield. Much of my work over the past twenty years has been with Ugandan colleagues to improve the vitamin and mineral content of their staple cooking bananas. It was very clear that during their wet seasons, the impact of Black Sigatoka on the harvest was to have a major effect on food security, as it still does. Can you imagine the psychological and physical desperation when, due to reasons out of your control, you lose half your food from the few acres you have to feed your family to a disease, and lose any opportunity to sell small surpluses in the local market? In contrast, the commercial export-oriented Cavendish plantations I have visited in South and Central America, and to a lesser extent in the Philippines, have shown little sign of the disease. Cavendish is highly susceptible to Black Sigatoka but constant spraying of fungicide keeps the disease at bay.

The 1960s ushered in Black Sigatoka but there was worse to come. In the 1990s Cavendish plants were reportedly dying in Taiwan with symptoms horribly reminiscent of Panama Disease of Gros Michel,[30] and by the turn of the century the same disease was found in Malaysia, Indonesia, China and Australia.[31] It was indeed Panama Disease, but not the strain that caused the pandemic in the Americas. This one was called Panama Disease tropical race 4, now commonly known as TR4, and it didn't evolve from race 1. It was an entirely 'new' type, though it has probably been around, co-evolving with bananas in some parts of the centres of origin in Asia, again for eons. In many ways, it is very similar to its predecessor in that it lives in the soil and infects bananas through their roots, ultimately killing susceptible varieties. The big difference however is that it infected and killed Cavendish.

And it does so very quickly; often there are only three to four weeks from the first signs of a yellowing leaf to full-blown wilt and ultimately death.

The Panama Disease TR4 pandemic had started. It's a slow-moving pandemic as it depends on the movement of the fungus, Foc, which itself depends on the movement of infected plants, soil and water. The soil can be attached to plants, boots, clothing or equipment and the fungus in irrigation or flood water. It is relatively easy for the fungus to move locally and even regionally. Between 2000 and 2015, it made its way into the first Cavendish exporting country, the Philippines,[32] as well as Pakistan and the Middle East. It soon made a big jump into Africa, specifically Mozambique, which was developing a new Cavendish export operation. These new outbreaks are always caused unwittingly by humans.

The disease continued its relentless spread across Asia and the Middle East, with one outbreak demonstrating just how easily it spreads. Somehow, Panama Disease TR4 made its way into the Eden Project in the United Kingdom. The Eden Project, now a major tourist attraction, was established in a china clay pit in Cornwall and consists of huge domes housing a myriad of exotic plants, including bananas: unfortunately Panama Disease

TR4 infected bananas.[33] The Eden Project is visited by more than 700,000 people per year, all walking through the domes on paths that are often wet. Efforts have been made to eradicate the fungus at the project. We can only hope this hasn't or doesn't become a source for the spread of the disease outside the UK.

The Cavendish producers and exporters in the Americas were aware of Panama Disease TR4 but not overly concerned. The disease hadn't been recorded in the western hemisphere and, while the Philippines was affected, they accounted for only 15 per cent of exports. That all changed in 2019. Panama Disease TR4 was identified in Cavendish plantations in the Santa Marta region of Colombia and in August of that year, the Colombian government declared a national emergency.[34] It takes a number of years after the arrival of the fungus in a plantation for the first symptoms of the disease to become evident, which unfortunately means it can significantly spread before any action is taken.

Within three years, the disease had been recorded in Peru and Venezuela.[34] Ecuador, the world's largest exporter, shares land borders with both Peru and Colombia. That country is now taking the threat extremely seriously. Relatively recently, I was invited to Ecuador to present at a conference specifically on Panama Disease TR4. On arrival from Australia, a TR4-affected country, having gone through customs, I was met by a number of plant quarantine officials who went through all my luggage very, very carefully, particularly inspecting my shoes. This was all filmed; I'm not sure who the target audience was. Needless to say, I am paranoid about ensuring there cannot be any fungus attached to my clothing or shoes when travelling.

Unfortunately, history would suggest the appearance of Panama Disease TR4 was only a matter of time, and this has been proven true as Ecuador recorded the disease in the south of the country in September 2025.[35] Despite very little border control of plant material being carried out between neighbouring countries, no other African country, other than Mozambique, has yet to record the disease.

The disease can move quickly in a newly infected plantation and once the first infected plant has been identified, a plantation can be close to wiped out within a few years. It is often difficult to find accurate information on the impact of TR4 within regions, either because the data has not been collected or because often there is an effort to downplay the effect the disease is having on the region. However, we have been able to collect real numbers over multiple years in one plantation, and from that it is not difficult to extrapolate to the broader industry.

Panama Disease TR4 has been present in the Northern Territory of Australia since the 1990s.[36] Known as the Top End, this is a huge territory but with a very sparse population, nearly 1.5 million square kilometres (approximately 600,000 square miles) with less than 250,000 people. Darwin, the capital city in the wet tropics, has a population of around 140,000 and is quite remote from any other significantly populated city or town. In the 1990s, there was a small local but thriving banana industry supplying Darwin and a couple of other regional towns. TR4 arrived some time in the 1990s. There has never been an explanation as to how it was introduced but Indonesia is geographically very close in a northerly direction and is presumed to be a likely initial source of the disease. It moved rapidly around the district until by ten years ago there was only one commercial plantation left. This is the plantation where we have our field trials. The trial site is about one hour by road from Darwin and has a variety of local and introduced inhabitants including freshwater crocodiles, magpie geese, snakes and buffalo.

Commercial Cavendish plantations generally produce fruit for in excess of seven years before replanting. Replanting is an expensive operation as it requires clearing the land of the current plants, allowing a fallow period, preparing the land for planting, purchase and planting of new plants, and finally waiting twelve months to harvest. There are smallholder plantations in the higher elevations of Uganda which have not been replanted for more than a hundred years. However, because of the impact of Panama Disease TR4, the Darwin plantation is replanted every one and a half to two years. This is only economically viable because transporting fruit to Darwin from outside the Northern Territory is very expensive. We planted our second, larger field trial in January 2018.[37] The first of the susceptible Cavendish plants were showing TR4 symptoms with six months of planting, although only a small percentage were affected. After six crops, approximately three and a half years later, nearly 75 per cent of the plants were either dead or infected. (More on this later.) The fungus is essentially in that soil for ever. Growing Cavendish as an annual crop is only viable in the Northern Territory because of the particular set of circumstances, so unless the western consumer is willing to pay well in excess of double the current price, this is not a strategy for the future. A group in the Netherlands, Neder Banaan, have grown bananas in a glasshouse and harvested the fruit, but this is more of an experimental exercise.[38] The actual production costs would be prohibitive and environmentally unsustainable.

A number of us made predictions twenty-five years ago that Panama Disease TR4 would become a 'superstar' disease, and that is coming to fruition. The media have had a field day with headlines such as 'Bananageddon' and 'The extinction of bananas', with the front cover of *New Scientist* in 2003 stating 'The last days of the banana: the world's favourite fruit is about to disappear'.[39] It is the 1950s revisited but with two big differences: there is no 'off the shelf' replacement for Cavendish as there was with Gros Michel, while Black Sigatoka has had an extensive impact in both commercial and smallholder plantations. What hasn't changed is that the vast majority of the media reports are related to the drama of the survival of Cavendish, the supermarket banana, for the world's western markets: the 15 per cent of the global crop as opposed to the 85 per cent consumed in developing countries.

Back in 1923, Frank Silver wrote a song about a Greek grocer in New York titled 'Yes, We Have No Bananas', certainly referring to the lack of bananas because of Panama Disease race 1.[40] If we don't find a way to control Panama Disease TR4, we may be hearing a new recording of that song.

The control of a new disease goes through a relatively standard evolution, whether it is a disease of humans, animals or plants. It starts with attempts at *exclusion*, followed by *eradication*, followed by *minimisation* and finally *resistance*. An excellent example is COVID-19. Once identified in China, many countries attempted to exclude the virus by preventing people coming in from China. That clearly didn't work, so attempts were made at eradication, which included the isolation of infected individuals and 'lockdown', dramatically reducing the area and times within which people could move. There was no eradication but, during that period, enough information had been gathered to ease those restrictions and move to minimisation which included social distancing, the wearing of masks, hand sanitation and isolation during infection, supported by inexpensive diagnostics. The two major diagnostics were lateral flow devices for 'home' use with nasal swabs, and PCR tests, the more accurate lab-based diagnostic. Despite popular thought, PCR, which stands for polymerase chain reaction (a good pub quiz question) wasn't invented for COVID but has been around since the 1980s and is widely used in research as well as diagnostics. Importantly, minimisation bought enough time to develop resistance: the vaccines. Unlike those for smallpox and, with more international cooperation, potentially polio, the vaccines in their present form will not lead to eradication but are undergoing continuous improvement to cope with a continuously evolving virus.

Consider Panama Disease, starting with race 1. There was no attempted exclusion as the fungus and the disease it caused had effectively been spread to all major production areas before it was even recognised as a disease. There were virtually no attempts at eradication. Minimisation was the major response for the export plantation growers. This included flooding of the fields and constantly moving to new lands. Finally, they turned to resistance: Cavendish. Interestingly, the smallholders and subsistence farmers of Asia, where the disease has been around for considerably longer, have coped by not growing monocultures of a single banana variety as well as by growing bananas in the company of many different food plants.

It has been around a hundred years since Panama Disease race 1 was widely recognised as having a real impact on banana export production in the Americas, and now we are now in a very similar situation. Panama Disease TR4 is widely recognised today as an existential threat to Cavendish. And we are going through the same evolution of control. There are major attempts at exclusion. We do now have a far greater understanding of the disease and the fungus that causes it. We understand how it moves locally and internationally and have implemented quite extensive biosecurity measures including advanced diagnostics, and local and international quarantine. Unfortunately, the fungus and the disease continue to spread. How did it move intercontinentally, first to Colombia, then Peru then Venezuela, and now Ecuador? The answer is via humans. Even with the best quarantine in place, it is almost certain to keep spreading; to Honduras, to Costa Rica and to the Caribbean.

We also now recognise that eradication is virtually impossible, so most 'infected' countries have implemented minimisation procedures. In the most severely affected export country, the Philippines, land clearing for new plantations is common and planting only suckers or plants that are known to be free of disease is possible.[41] Local strategies also include procedures for disinfecting shoes and clothing, implements and vehicles. These procedures are becoming standard practice. Other strategies include planting selections of Cavendish which seem to get infected more slowly, treating the soil with beneficial micro-organisms and growing other crops such as Chinese leeks between the rows, which appears to result in fewer infections.

These are all stop-gap measures; Cavendish doomsday is coming. We have to move to resistance. Unfortunately, there are no resistant varieties available that could easily replace Cavendish. This time one has to be created; there is time to do so before there is serious impact on global production.

Of course, we can't ignore the other major disease, Black Sigatoka. Apart from a very special case in Australia, Black Sigatoka skipped the exclusion and eradication strategies and has gone straight to minimisation through chemicals or dry climate production. Ideally, the new banana 'creation' would be resistant to both Panama Disease TR4 and Black Sigatoka.

Prior to 1990, there were few options for creating new plant varieties. By far the most common was conventional breeding, taking the pollen of the chosen male donor plant onto the stigma of the chosen female recipient plant, triggering fertilisation of the ovule. Sexual reproduction is remarkably similar across all kingdoms. Conventional breeding of bananas was and continues to be challenging, primarily because the vast majority of domesticated and therefore desirable varieties are close to sterile, resulting in seedless fruit of great benefit to the consumer, but offering a monumental challenge for the plant breeder.

Obviously with some level of forethought, United Fruit, now Chiquita, established a banana breeding programme in Honduras in 1959.[42] This was after the Gros Michel Panama Disease pandemic but just before the global march of Black Sigatoka, and well before the advent of Panama Disease TR4. In 1969, Phil Rowe, an American from Arkansas, was appointed chief breeder.[43] I first met Phil when I was in my very early days of working on banana diseases. He had a wonderful Southern drawl and his presentations were fascinating, but his approach to breeding new banana varieties was on another level altogether. His breeding strategy was to identify disease-resistant diploid fertile bananas, collect pollen from them and attempt to fertilise literally tens of thousands of female flowers on popular varieties, which were of course close to sterile. This was a huge operation every year and was supported by a very large workforce. Timing of course was critical, as the pollen had to be fresh and the female flowers receptive. Every year, Phil and his team would harvest in excess of 20,000 bunches when fully ripe and the flesh of all these fruits were forced through coarse sieves into 44-gallon drums. The result: just a few seeds per year, which were treated as if they were pure gold.

During this time, when Phil was toiling away fertilising flowers and collecting seed, Chiquita donated the breeding programme to Honduras and it became the *Fundación Hondureña de Investigación Agrícola*, better known as FHIA. Phil was screening his newly generated hybrids for three main characteristics: resistance to disease (particularly Black Sigatoka), bunch size and fruit characteristics. From the programme, a number of hybrids were released. They included dessert and cooking types; some were dwarf while others, such as FHIA-25, were enormous, and most had multiple disease resistances.

Perhaps the most successful was FHIA-1, better known as Goldfinger.[44] It has excellent resistance to Black Sigatoka as well as the less severe Yellow

Sigatoka and Panama Diseases and produces big bunches, but unfortunately it has faced little consumer acceptance. It was released in Australia in 1995 to much fanfare; by 2009, the supermarkets stopped selling it and now it accounts for less than 1 per cent of bananas consumed in that country. The taste and texture didn't appeal to consumers and farmers found it more difficult to produce. This wasn't the first conventionally bred banana release. Way back in the 1920s, a Trinidad breeding programme released a banana called IC2. While it gained some traction in the American market, it was probably not completely resistant to Panama Disease race 1.

Tragically, Phil Rowe died in 2001 but his legacy lives on. He demonstrated that it was possible to breed bananas with high levels of multiple disease resistance, which had not been achieved previously. The FHIA breeding programme continues but with a more nuanced strategy, and is now supported by another of the multinationals, Dole. One of the key messages for all others attempting to develop a new variety was that fruit quality is not negotiable. The western consumer has little to no interest in disease control: the fruit must be brilliant yellow, curved, and taste like Cavendish.

After the first releases from the FHIA breeding programme, there were some remarkable developments that dramatically changed banana breeding strategies, including how to efficiently generate new triploid seedlings. Most of the FHIA hybrids were tetraploids, meaning that each cell had four sets of chromosomes. As I explained earlier, domesticated varieties are invariably triploid with three sets of chromosomes, which almost certainly accounts for their sterility. The strategy most banana breeders are now following is to generate new triploid varieties, in reality trying to reproduce what Nature's breeding programme has generated. To do this, the female parent is a diploid that has been treated with a particular chemical that stimulates doubling of chromosomes, so that the treated plants become tetraploid, or rather, double diploids. Many of these have a reasonable level of fertility. These fertile females are then crossed with pollen from the same or another fertile diploid, and the majority of the resulting progeny are triploid and sterile. However, these crosses yield vastly more seed than the Phil Rowe approach. This has been a major advance.

The other major advance has been the genomics revolution, in short the ability to clone DNA, to sequence this DNA from single genes to entire genomes, to identify genes and gene function, and then, through advanced genetic techniques. to manipulate and modify genes and ultimately to artificially move genes.

The genomes of a number of different bananas have been completely sequenced.[45] A banana has approximately 35,000 genes. In comparison, we humans have about 19,900 genes. The genomes of the two banana species, *Musa acuminata* and *Musa balbisiana*, multiple genomes of Cavendish as well as the genomes of the three wild subspecies that contributed to the Cavendish genome, namely *malaccensis*, *zebrina* and *banksii*, have all been sequenced. These days that's the easy bit; we can have a banana genome sequenced in a few weeks and for a few thousand dollars. Far more complex and challenging is determining what all these genes do. The specific functions of many of the genes are unknown and the influence of the different versions, alleles, is even less well understood. However, while the precise identity of the ever-important disease resistance genes is not yet known, we do know the general location of some of these genes. This requires some explanation. Bananas have eleven different chromosomes. Triploid bananas such as Cavendish have three copies of each chromosome, thus thirty-three in total in each cell. Together the eleven banana chromosomes are made up of DNA consisting of about 580 million base pairs or 'letters'. It is possible to think of chromosomes as chapters of a book, the genetic book or perhaps the instruction manual. In English, we use an alphabet of twenty-six letters while DNA uses just four 'letters', A, C, G and T. All the instructions in the DNA book of a species and an individual are written down as a sequence of these four 'letters', or more precisely bases.

Thus, we can easily read and understand the instruction 'Build the wall with blocks'. A simple readable sequence of letters. However, here is an actual example of a gene instruction:

```
agccctccag gacaggctgc atcagaagag gccatcaagc aggtctgttc caagggcctt

tgcgtcaggt gggctcagga ttccagggtg gctggacccc aggccccagc tctgcagcag

ggaggacgtg gctgggctcg tgaagcatgt gggggtgagc ccagggggccc caaggcaggg

cacctggcct tcagcctgcc tcagccctgc ctgtctccca gatcactgtc cttctgccat

ggccctgtgg atgcgcctcc tgcccctgct ggcgctgctg gccctctggg gacctgaccc

agccgcagcc tttgtgaacc aacacctgtg cggctcacac ctggtggaag ctctctacct

agtgtgcggg gaacgaggct tcttctacac acccaagacc cgccgggagg cagaggacct

gcagggtgag ccaactgccc attgctgccc ctggccgccc ccagccaccc cctgctcctg

gcgctcccac ccagcatggg cagaaggggg caggaggctg ccacccagca gggggtcagg

tgcacttttt taaaaagaag ttctcttggt cacgtcctaa aagtgaccag ctccctgtgg
```

cccagtcaga atctcagcct gaggacggtg ttggcttcgg cagccccgag atacatcaga

gggtgggcac gctcctccct ccactcgccc ctcaaacaaa tgccccgcag cccatttctc

caccctcatt tgatgaccgc agattcaagt gttttgttaa gtaaagtcct gggtgacctg

gggtcacagg gtgccccacg ctgcctgcct ctgggcgaac accccatcac gcccggagga

gggcgtggct gcctgcctga gtgggccaga cccctgtcgc caggcctcac ggcagctcca

tagtcaggag atggggaaga tgctggggac aggccctggg gagaagtact gggatcacct

gttcaggctc ccactgtgac gctgccccgg ggcgggggaa ggaggtggga catgtgggcg

ttggggcctg taggtccaca cccagtgtgg gtgaccctcc ctctaacctg ggtccagccc

ggctggagat gggtgggagt gcgacctagg gctggcgggc aggcgggcac tgtgtctccc

tgactgtgtc ctcctgtgtc cctctgcctc gccgctgttc cggaacctgc tctgcgcggc

acgtcctggc agtggggcag gtggagctgg gcgggggccc tggtgcaggc agcctgcagc

ccttggccct ggaggggtcc ctgcagaagc gtggcattgt ggaacaatgc tgtaccagca

tctgctccct ctaccagctg gagaactact gcaactagac gcagcccgca ggcagcccca

caccccgccgc ctcctgcacc gagagagatg gaataaagcc cttgaaccag

It is impossible to understand unless you use the genetic dictionary, Genbank,[46] where you would discover that this is the DNA sequence or 'instruction' of the human insulin gene.

I should also mention that changing one letter can have a dramatic effect on the 'instruction'. For instance: change one letter in 'Build the wall with *b*locks' to 'Build the wall with *a*locks', changing 'b' to 'a', and now the instruction is wrong and nonsense. It is the same with DNA. A change, or in other words a mutation, in just one or a few letters can completely change the instruction, for better or for worse. This happens naturally at a very low, often undetectable rate between generations, but can also be induced by irradiation and more recently by gene editing. Mutations, of course, are one of the key elements driving evolution.

We now have the sequence of the approximately 580 million letters making up bananas, but we don't yet understand the functions of most of the genes/instructions.[44] What is known are the approximate locations of some of the resistance genes. This discovery has been achieved through a process called gene mapping. For instance, the Cavendish gene for resistance to Panama Disease race 1 is towards one end of Chromosome 10 and the gene for resistance to Panama Disease TR4 in one *malaccensis* plant is on Chromosome 3. These locations are known as markers and are being used by plant breeders to track the inheritance of these markers, and therefore resistances, from parents to offspring, even though the exact resistance gene hasn't been identified.

So, armed with genomes, location markers, efficient generation of triploids and sources of genes for characteristics, traits, such as resistance, conventional banana breeding has entered an exciting new era. A number of programmes are attempting to recreate Cavendish or something similar but with resistances. The selection of parents is obviously crucial and is affected by having both desirable traits and an accessible genomic sequence. Breeders guard the identity of the parents they use for obvious reasons. However, some of them are well known and have interesting names such as Calcutta 4 and Cultivar Rose. These are both fertile diploids. Phil Rowe used Calcutta 4 extensively in developing his diploid male parents and, notably, it was also used in the very early breeding programmes in Trinidad. A wild-seeded/fertile diploid probably originating in the vicinity of Myanmar, it is a *Musa acuminata* subspecies *burmacoides*, and has excellent disease resistances.

As I mentioned previously, the major genetic contributor to our favourite varieties is *Musa acuminata*, but many of the lesser-known varieties, including all of the plantains, have up to a third of their genes from *Musa balbisiana*, the other major banana species. This species is the more robust and often has some drought tolerance as well as disease resistances. It was being used in some breeding programmes but it came with an unpleasant and completely unforeseen surprise. Hidden away in the *balbisiana* genomic DNA are fragments of a virus called banana streak virus.[47] Sometime in prehistory, a progenitor of bananas was infected with this virus and fragments of the virus became part of the genome. The unpleasant surprise was that, when *balbisiana* was used as a parent, these individual fragments recombined to produce a fully functioning infectious virus in the resulting offspring. At the time, Frederick (Fred) Bakry[48] was the lead breeder at CIRAD, the French research and cooperation organisation focusing on international sustainable agricultural development. He worked out that *balbisiana* was the likely cause and, subsequently, it was dumped as a parent despite some of its desirable traits. I should mention that *Musa acuminata* also has fragments of banana streak virus in its genomic DNA, but they seem unable to recombine to produce an infectious virus.

CIRAD has been the early leader in combining diploid based breeding and genomic analysis with Fred Bakry as the breeder and Angelique D'Hont leading the first team to sequence a banana genome.[49] With a breeding programme based primarily in Guadeloupe, Fred and the supporting genomics team have produced a large number of new hybrids. They were probably the first to use the strategy of crossing a doubled diploid with a diploid. Their efforts have not been confined to a disease-resistant replacement for Cavendish, as they are committed to supporting local domestic production and consumption as well as the export market. A key target has been resistance to Black Sigatoka, as that disease has had a considerable impact in Guadeloupe and other islands of the Caribbean. This programme is the most advanced of all the 'new generation' breeding programmes and lines with resistance to Black Sigatoka have been identified in Guadeloupe. A number of these have been screened in the Northern Territory of Australia for resistance to Panama Disease TR4. As far as I am aware, the results have been very encouraging, but as yet a disease-resistant Cavendish lookalike has not been identified.[50]

Others have joined the race to develop the next-generation Cavendish replacement. Chiquita has re-entered the fray through a company called Yelloway[51] with the stated aim of 'recreating' Cavendish but with resistance to both Black Sigatoka and Panama Disease. Yelloway has invested heavily in genomic analysis through shareholder Keygene, thus identifying the likely progenitors of Cavendish and using these in their breeding programme. The breeding takes place in glasshouses in the Netherlands. Apparently, through 150 crosses, they have produced 32,000 seeds. Much of the initial resistance screening is done in the Netherlands while the field trials have taken place in the Philippines. In September 2024, Yelloway announced they had produced a prototype triploid banana, Yelloway-1, which was resistant to Panama Disease TR4 and had partial resistance to Black Sigatoka.[52]

Dole has also invested in developing both genetically modified Cavendish through a partnership with ELO Life Systems in the USA and a conventionally bred Cavendish replacement with FHIA.[53] Some of the more recent FHIA lines have been trialled for Panama Disease TR4 resistance in Australia.

Other conventional breeding programmes around the world include EMBRAPA in Brazil,[54] the National Research Centre for Banana in India[55] and at least one programme in China, the Institute of Fruit Tree Research. Based on the rapid increase in the number of papers published in international journals, China is investing heavily in banana research and development and the outcomes of that investment are clearly evident. In contrast, there is comparatively little banana R&D investment in the USA other than through international aid programmes.

The other large programme is headed by the International Institute for Tropical Agriculture, whose headquarters are in Ibadan, Nigeria.[56] The Breeding Better Bananas programme has many collaborating institutions from around the world and is focused on developing new varieties of cooking bananas, primarily for East Africa, rather than a Cavendish replacement.

Conventional breeding of bananas has come a long way. Starting from the 'one seed per thousand bunches' of Phil Rowe's programme, new strategies are producing literally thousands of hybrid seeds. However, significant challenges remain, the most important of which are the limited sources of multiple disease resistances and the inability to 'back-cross'. Backcrossing is a key strategy used by plant breeders to transfer beneficial genes while reducing the impact of unwanted genes. It's a relatively simple concept and technique. When two diploid parents are crossed, then the offspring has an equal mixture of genes from both parents. As an example, let us consider a situation where one parent produces wonderful fruit on big bunches but is also highly susceptible to diseases, whereas the other parent is resistant to diseases but has poor-quality fruit and small bunches. With this in mind, the first-generation offspring individuals are selected because they are disease resistant, but at the same time they will almost certainly have also inherited some of the undesirable traits from the resistant parent. These first-generation individuals are then crossed back to the original 'desirable fruit' parent. The resulting second-generation offspring can again be selected for resistance and at the same time will have 75 per cent of the genes of the desirable parent. This process can be repeated over and over again, with each generation reducing the influence of the poor-quality original parent while retaining resistance. The problem for bananas is that the 'desirable fruit' offspring is sterile so there is no prospect for backcrossing. This means that the conventional breeder must find the best-quality parents and produce literally thousands of first-generation offspring to have a hope of getting that 'one in a thousand' banana with excellent disease resistance and wonderful, plentiful fruit.

A far less common or popular approach to generating new varieties is irradiation. This is where the reproductive tissue of a plant is treated with ionising radiation to damage the plant DNA and therefore, by chance, introducing changes or mutations in the DNA that may result in a new desired characteristic. This approach is being used in Australia in an attempt to improve the fruit quality and consumer acceptability of Goldfinger. It's still early days.

I should also mention a group of selections which are known as somaclonal variants. The best way to describe these is by way of an example. GCTCV 218 is a selection from Taiwan.[30] GCTCV stands for Giant Cavendish Tissue Culture Variant. It was selected because it appeared to have tolerance to Panama Disease TR4 in the field. It is widely grown in the Philippines but it doesn't yield as well as standard Cavendish and has some bunch abnormalities. In our field trial in Australia, it wasn't any more tolerant than standard Cavendish. There are quite a few other naturally occurring variants that have been selected and reported to have disease tolerance. At best, these are stop-gap varieties until real resistance is achieved.

While the field is dominated by efforts to solve the Cavendish problem, there are efforts to breed other banana types. Although export bananas represent only about 15 per cent of the global crop and are produced mainly in monocultures in South and Central America and the Philippines, a further 35 per cent of global production is also Cavendish. The fruit is extremely popular in the domestic markets of the two main producers, India and China, as well as in other non-exporting countries with big domestic markets. These include Brazil in South America as well as many countries of sub-Saharan Africa, and countries in South-east Asia such as Malaysia, Vietnam and Cambodia, though these latter countries are looking to get into the export market.

But what about the other 999 or more other varieties and landraces, the cooking bananas, the plantains and the vast array of other dessert bananas? Almost all of these have disease problems alongside other traits which limit their production or commercialisation.

The cost of developing a new crop variety is significant. The multinational marketers, including Chiquita, Dole and Fresh Del Monte, are at last investing considerable funds directed almost exclusively towards the supermarket product, which is understandable; that is their market. This has a spillover effect for countries where Cavendish forms a significant proportion of the domestic market, as any advances can be directly applied to domestic crops. In contrast, in the countries where other types of bananas are part of the culture and form an important part of the local diet, the funding is provided by national governments, and most significantly by international aid programmes or philanthropic organisations. The top six non-export banana producers are India, China, Uganda, Indonesia, Nigeria and Brazil. The governments of countries like India, China and

Brazil support agricultural research and development. However, other countries can be left behind, relying on funding from organisations in the developed world. For bananas, the Bill & Melinda Gates Foundation, USAID, the Rockefeller Foundation and the Sainsbury Foundation have been prominent as leading individual direct donors. The Consultative Group on International Agricultural Research is also significant, overseeing a portfolio of fifteen different agricultural research centres based primarily in Africa, Asia and South and Central America. Their two centres focusing on banana research and development are the International Institute of Tropical Agriculture (IITA), headquartered in Nigeria, and the Alliance of Biodiversity International and the International Center for Tropical Agriculture (CIAT), with its headquarters in Colombia.

Brazil is a standout when it comes to research and development for a predominantly domestic market. About 60 per cent of their production is Cavendish but around 30 per cent is Prata Ana or Brazilian Dwarf. The latter has small fruit and is very sweet, but is unfortunately susceptible to Panama Disease race 1 (and probably TR4, as well as Black Sigatoka). The other major bananas produced are plantains at around 7.5 per cent, and the Apple banana at 1.5

per cent. Brazilians love their Prata Ana and Apple bananas. EMBRAPA is using conventional breeding to generate a disease-resistant Prata Ana/Apple banana replacement and is already making good progress. Some of the EMBRAPA hybrids have been sent to Colombia to determine whether they are resistant to Panama Disease TR4, which at this time does not occur in Brazil. Interestingly, Prata Ana was the female parent of FHIA-1, the banana also known as Goldfinger. The resistance of Goldfinger to Panama Disease and Black Sigatoka comes from its male diploid parent, SH-3142, one of Phil Rowe's fertile-seeded diploids.

Conventional breeding of bananas is now at the point where large numbers of progeny can be generated. Disease-resistant new varieties have already been developed but as yet, there is nothing approaching a replacement for the Cavendish. It's a numbers game. In time, perhaps within a decade, there will be one or more conventionally bred disease-resistant dessert bananas suitable for the export market, but how close they will be to Cavendish is the great unknown.

The Next Generations

The act of deliberately crossing or hybridising two plants to generate a third, different plant dates back nearly three hundred years. It formed the basis of the huge seed industry of today. In contrast, the act of moving a single gene from one plant to another goes back only about forty years, but in those forty years the technology has rapidly developed, forming the umbrella of what is now known as the New Breeding Technologies. The most advanced and the most controversial is crop genetic modification, the process of adding new genes as DNA into a plant. Over 200 million hectares of farmland are now planted worldwide with genetically modified (GM) crops.

GM technology depends on a bacterium that occurs naturally in the soil, and unintentionally has provided us with a method of inserting new DNA into plants. It is truly amazing just how bacteria, viruses and fungi, collectively known as microbes, have provided the biotechnology methods that underpin much of our existence today. From the early exploitation of yeasts to make beer, wine and bread, and the first vaccines such as cowpox, through to the recombinant DNA revolution that now underpins advances in human health, it has been microbes that have provided most of the tools to move DNA, clone DNA, sequence DNA and manipulate DNA Unfortunately for us, it is also microbes that cause the vast majority of infectious diseases of humans, animals and plants. Give and take.

Agrobacterium tumefaciens, known in the industry as Agro, is a soil-inhabiting bacterium and a natural genetic engineer.[57] In nature, it infects a very wide range of plants and with a little coaxing, it can infect virtually any plant. It causes a disease called crown gall, and its trick is to invade a plant cell, injecting a piece of its DNA and some of its genes into the cell nucleus and into the plant's DNA. These genes essentially become additional 'plant genes'. The bacterial genes direct the cell to make sugars that only Agro can use, also inducing the cells to divide, resulting in tumour-like outgrowths called galls. It has been a remarkably successful strategy and there are hallmarks of its lifestyle way back in time. At some point, sweet potato must have been an early victim of Agro because today most sweet potatoes that we eat have genes that can be traced back to Agro.

While this proved a very smart strategy for a long time, scientists in the 1980s worked out that it was possible to replace the Agro genes that were transferred into the plant DNA with any gene of choice, and once the Agro had done its transfer job, it was possible to kill it off. The result: a genetically modified plant, otherwise known as a transformed or transgenic plant. Other methods of introducing new DNA into plants have been developed but none have the flexibility and broad applications of Agro-mediated transformation.

I decided back in the early 1990s that genetic modification provided us with a wonderful opportunity to genetically improve bananas. We decided our GM philosophy would be to take what Nature's massive breeding programme had provided and improve on it only where necessary. Rather than try to recreate Cavendish or Gros Michel or East African Highland bananas, we would attempt to add a few genes, preferably banana genes, to these very popular varieties to rectify specific susceptibilities and flaws or further enhance specific traits. These varieties have been stuck in a time warp because they are sterile. However, this did not mean that they should be discarded: with genetic modification, we deemed that we should be able to provide any necessary evolutionary boost or adaptation in the future.

I should stress that I made the decision to focus on developing GM bananas before any GM crops were grown commercially and before the almost instantaneous opposition to these crops. With hindsight, would I make the same decision today? Absolutely yes. At the time, only Rony Swennen at KU Leuven in Belgium was attempting to develop GM bananas.[58] He is now leading a major conventional breeding programme for Africa. Others joined the effort later.

It quickly became evident that the path to a GM banana was not going to be lined with myriad of enthusiastic supporters. However, at that time, we were focused on the scientific and technical challenges rather than the consumer acceptance challenges. These were to come later. First, we had to develop a way to genetically modify or transform bananas, particularly Cavendish – that is, a way of introducing new genes into the genome. Bananas are not a natural host of Agro so our first attempts were based on biolistics, and it worked. Biolistics involves coating tiny gold or tungsten particles in DNA and literally blasting these into cells. It is quite efficient but the DNA tends to get broken up into fragments, which is not ideal.

There were others around the world attempting to do the same thing and, for a period, the three main groups – ours at Queensland University of Technology (QUT), Rony Swennen's in Belgium and Charlie Arntzen's group at Cornell University in the USA – were funded by the World Bank as part of a large international initiative on banana improvement. Just before the year 2003, we had finally developed a method for Agro to modify Cavendish.[59] Charlie Arntzen's group also developed a method, though it was less efficient.[60] We could now create GM bananas at will.

The process involves generating what we call embryogenic cell suspensions. We collect banana flowers and cut out the immature male flowers. These are genetically identical to the mother plant from which they came. The flowers are placed in sterile culture on a medium that contains nutrients as well as plant hormones to induce them to form somatic embryos (which are embryos identical to the tissue they came from). At this stage, the medium is made of agar, which forms a very solid gel into which the nutrients and hormones are mixed. The embryos are omnipotent and from a single cell a whole plant can be regenerated that is identical to the plant from which the immature male flowers were taken. These cells are the equivalent of human stem cells, which provide pathways to cure human diseases that were previously incurable. And, of course, there was Dolly the cloned sheep. Dolly was grown from stem cells and so was an exact copy of her parent.

We then mix these cells in a liquid with Agro, which carries the genes we want to insert in the cells' genome; Agro does the transfer, we kill off the Agro with antibiotics and then, by manipulating the plant hormones, induce the transformed cell to multiply and form clumps. When large enough, the clumps become embryos and germinate, producing roots and shoots. During this period, the process all takes place in sterile culture. While the leaves are green, photosynthesis is providing very little if any energy. It all comes from the sugars in the media. When the plantlets are big enough, we 'cut the umbilical cord' and transfer the plantlets to soil where they have to feed by themselves. The whole process requires some considerable time. It takes about nine months to generate the embryogenic cells, which is done on a yearly basis. It then takes a further nine to twelve months from the time we insert the genes into a single cell until we have a fully formed plant in soil.

The 2000s became both exciting and ever more challenging. Until about 2004, we were funded by an Australian government scheme with matching funding from the Australian Banana Growers Council. However, this organisation became increasingly concerned about consumer attitudes to GM and understandably took a conservative stance, cancelling the funding for our research. It wasn't reinstated until very recently, almost twenty years later.

Panama Disease TR4 had arrived in the Northern Territory of Australia and was making the production of bananas there very difficult. The region accounted for less than 1 per cent of Australia's production; there was national concern about the appearance of the disease but it was not excessive. Despite this, we made the decision to focus our efforts on developing a GM Cavendish resistant to Panama Disease TR4. This was before TR4 became a 'superstar disease'.

We needed a gene that would confer resistance to TR4. Banana genomics wasn't very advanced in those days but we were able to do some rudimentary searching for possible TR4 resistance genes and we screened a number of different bananas without finding anything promising. Then, two crucial encounters with a very talented, somewhat eccentric American banana plant pathologist called Ivan Buddenhagen changed everything.

Ivan was an inveterate collector of specimens related to plant diseases, particularly banana diseases. On an expedition to South-east Asia, he came across a field of Cavendish that had been decimated by TR4 but, in the same field, there were some banana plants growing tall and healthy. They weren't Cavendish but *Musa acuminata* subspecies *malaccensis*, a wild diploid banana with fruit full of seeds. He collected those seeds and sent them to colleagues in Australia, who germinated at least six of them and tested them for resistance to Panama Disease. Three were resistant and three were susceptible. This was the perfect set to look for the elusive gene. And, lo and behold, one of our PhD students identified a very promising candidate gene which we have called RGA2.[61] We very quickly transferred that gene into Cavendish, making one significant modification. An essential part of a gene is the promoter, which is essentially an on/off and volume switch for the gene. Most plant-resistance genes are on all the time but at a very low level. We changed the native

RGA2 promoter for one with a higher volume; a critical decision in hindsight.

We generated seven different genetically modified plants, each carrying the new RGA2 gene, and had those plants growing in our PC2 – Physical Containment 2 – glasshouse. Much of our work is highly regulated. To develop GM plants, we must get approval first to make the plants, then to test them in a glasshouse, and then again to test them in the field. For the glasshouse test, the GM plants in Australia must be confined to PC2 glasshouses. In addition, at the time we made our first GM plants with the RGA2 gene, TR4 hadn't been found in Queensland, where both we and 95 per cent of the Australian banana industry are located. The Queensland state government understandably wouldn't allow us to bring Foc, the TR4 fungus, into Queensland to use in our PC2 glasshouse. TR4 did occur in the Northern Territory but there were no PC2 glasshouses there. We were stuck.

Enter Ivan Buddenhagen for the second, crucial time and the start of a long and productive collaboration with an innovative banana grower, Robert Borsato. Robert had plantations in Queensland but had expanded to develop a large plantation in the Northern Territory. Very quickly this new plantation was affected by Panama Disease TR4. Robert was desperate to find a solution and searched the world for the best banana plant pathologist. He found Ivan, who came to the Northern Territory, went through a number of possible approaches to at best ameliorate the impact of Panama Disease TR4 and then said, 'We need to go to Brisbane and visit James Dale.' Ivan had seen a number of my presentations at conferences so he was well aware of what we were doing. At our first meeting in Brisbane, Robert volunteered his plantation for a field trial and so we went straight to the field without the usual intermediate glasshouse assessment. We were back in business!

From there, we made rapid progress. We planted our first small-scale field trial on Robert's plantation at Lambell's Lagoon, close to the town of Humpty Doo in the Northern Territory, in January 2012. The region is populated by a variety of dangerous animals, and sharing such a site with them can be tricky: snakes have an unfortunate habit of coiling up in the bunches, which can make life interesting during harvest, and we are also very aware that young saltwater crocodiles find irrigation canals an attractive environment. After three years, four of the lines showed excellent resistance to Panama

Disease TR4, one having no infected plants at all, while the normal Cavendish were rapidly dying.[62] Subsequently, in January 2018, we planted a much larger trial, one which continues today. The results have been remarkable. One line, called QCAV-4, has outperformed the other three resistant lines and after more than three years, the infection rate is only 4 per cent, whereas for the standard non-GM Cavendish, most of the plants are dead and the infection rate was 80 per cent. Very importantly, the yield was not affected and the fruit looks, feels and tastes exactly as a well-grown Cavendish should.[63]

In 2023, we took the next major step, applying for 'deregulation' of QCAV-4 in Australia which would allow it to be grown commercially and sold in Australian markets and shops. This required applications to two different regulatory organisations: the Office of the Gene Technology Regulator (OGTR) [64] which regulates cultivation of GM crops, and Food Standards Australia New Zealand (FSANZ)[65] which regulates food for human consumption. We essentially had to demonstrate that the QCAV-4 was as safe as a conventional banana. This was relatively straightforward. Many of the elements we utilised had been used numerous times in other deregulated GM crops such as cotton, soybeans and canola. The gene we used, RGA2, is a banana gene and seems to be present in the genomes of most bananas, including Cavendish. However, in Cavendish, the gene doesn't seem to work. And because Cavendish is sterile, there is no chance that the RGA2 gene will spread from QCAV-4 to other non-GM bananas.

By April 2024, we received approval. Australian farmers could now grow QCAV-4 and sell the fruit in markets and shops. This was a huge step forward: it is the first GM banana to be approved for growing and eating anywhere in the world, and the first GM fresh fruit in Australia. GM bananas were a reality after thirty years.

The most important outcome is that a Cavendish banana is now available that is resistant to Panama Disease TR4 and yields as well as its non-GM 'parent'. We don't know – and won't know for some considerable time – whether an equivalent new variety can be produced through conventional breeding programmes.

Developing a new banana variety in this way is not a quick process: from my experience, it takes at least nine years from starting to finally having that variety ready for the market. It is at this point that the 'commercialisation phase' commences with the applications for Plant Breeders Rights (PBR).

Let me use QCAV-4 as an example. To get PBR on QCAV-4, we had to show it was physically different to any other banana variety. The challenge for us was that QCAV-4 looks identical to its parent except that it is resistant to Panama Disease TR4. We therefore had to show the 'examiner' the field trial where QCAV-4 was thriving, and prove its non-GM parent was dead or dying. We have already been granted Plant Breeders Rights in Australia and in the USA.[66]

As with many other regulations, the application process and criteria varies from country to country, which means that we have to prepare different applications for different countries; a time-consuming and frustrating process as the fundamental information is the same. To conform to quarantine regulations, bananas are almost always moved internationally in sterile media and are usually accompanied with a certificate of pest and disease freedom from the exporting country. While complex and time-consuming, this process is reasonably effective for excluding unwanted pests and diseases. However, we are well aware that there are many instances of illegal importation of plants and this would be an easy route for the introduction of Panama Disease.

For QCAV-4, we are in the process of getting approval to transfer plants to the Philippines for field trialling. As these plants are genetically modified, there is another level of approval needed, plus approval to conduct what is known as a confined field trial. Confined field trials are highly regulated, but the regulations again differ from country to country. In Australia, a simple wire fence with a sign is sufficient. In Uganda, we needed a 2 metre high chain wire fence, presumably to ensure the plants weren't stolen. Unfortunately, someone stole the fence instead! The approvals for the Philippines field trials have been granted.

In this world of rapid technological advances, there will always be new technologies being developed. While we were beavering away developing genetically modified bananas, a technology called gene editing – and particularly the CRISPR version ('clustered regularly interspaced short palindromic repeats') – was being developed. There is a major difference between gene editing and genetic modification. With genetic modification in plants, we insert additional DNA, usually a gene, into the genome.[67] For instance, we inserted the RGA2 gene into the Cavendish genome to make QCAV-4, resulting in disease resistance, and we added a gene called phytoene synthase to Cavendish and East African Highland bananas to make golden fruit. With gene editing, we change the DNA in the genome rather than adding it. CRISPR acts like scissors and cuts the DNA only at the site or sequence it is programmed to cut. This programming is highly specific and can be altered to cut virtually any sequence. For instance, CRISPR can be programmed to make just one cut in the 580 million bases (or letters) of the banana genome, at a designated targeted site. We can introduce the CRISPR mechanism into a cell which includes an enzyme called Cas which actually does the cutting; the DNA gets cut and the cell then repairs it while usually removing one or more bases. We can go back to my analogy from earlier, when I used the example instruction of: 'Build the wall with *blocks*'. After a CRISPR/Cas cut and cell repair, the sentence/instruction could now read 'Build the wall with *locks*' where one letter has been deleted, or 'Build the wall with *ocks*' where two letters have been deleted. Now the instructions are nonsense.

In plants, this is the simplest and most common application of CRISPR, that is, to knock out the function of target genes. The instruction of the gene is meaningless and so the function of that gene is destroyed.

The next level of CRISPR is to change specific letters so as to change rather than destroy the instruction, as in: 'Build the wall with *bricks*' where the two letters *l* and *o* of 'blocks' have been replaced by the two letters *r* and *i* for 'bricks'. It is now a valid but different instruction. This is already being done in plants but is a more challenging process. In animal cells, it is possible to replace an entire gene. This hasn't been done in plants yet, but it will come.

Why is this technology so important? A major reason is regulation. As I have said, genetic modification is highly regulated and is not even permitted in some countries, and the product in many instances has to be labelled as GM. This results in huge costs and consumer wariness at best. Already the regulatory authorities in some countries have determined that many of the applications of CRISPR, such as gene knockouts, can and do occur naturally and therefore they should not be regulated or labelled. This is the situation in the USA, Japan and England, and now in Australia.[68] Even the European Union is in the advanced stages of considering how they might regulate gene-edited products as non-GM.[69]

CRISPR editing has another large but non-scientific advantage. New technologies are invariably patented, which means that for others to use the technology, they must get a licence. In the early days of genetic modification, the major patents were held by the large multinational ag biotechnology companies including Monsanto, Bayer (which now owns Monsanto), Syngenta and BASF. For small organisations, it was nearly impossible to get a licence for these patents. Time passes and now most of the patents have expired as they have a finite life. Of course, CRISPR is also patented – in fact, there are many patents – but this time round a company called Corteva has acquired the rights to much of the technology in plants and most importantly is willing to negotiate licences to use the technology to others.[70] My university, QUT, already has an R&D licence. This has provided smaller organisations with the confidence and impetus to develop gene-edited crops.

CRISPR has already been used to edit the genome of bananas and three groups, including my group in Australia,[71] Leena Tripathi's group in Kenya[72] and Sidat Tiwari's group in India[73] have published these results. Two private companies, Tropic Biosciences in the UK and ELO Life Systems in the USA, are also developing edited bananas. It is now possible to gene edit bananas without any new DNA being inserted into the banana genome. In an increasing number of countries, these edited bananas would not be considered to be GM and would not be regulated. This has multiple implications: development time will be quicker because regulatory approvals are usually very drawn-out processes, the development costs will be lower because the time burden of compliance will be reduced, market access will be greater, particularly if the EU adopts their current draft proposals, and greater consumer acceptance is likely as the fruit will not have the GM label.

The early targets are the ones where we already have a good understanding of which genes to edit, particularly doing simple knockouts of function. Perhaps one of the more advanced programmes is being conducted by Tropic Biosciences.[74] They have reported in the media that they have generated edited Cavendish whose fruit has been altered to delay ripening and thus prolong shelf life. This is a relatively easy knockout edit. A Cavendish with a long shelf life, if commercialised and popular in Western countries, could have a dramatic impact on the global export market, and as is normally the case, small producers would likely be those most seriously affected.

Of course, the big target is resistance to Panama Disease TR4. There are two main editing approaches. One is to make the plants more resistant by targeting banana resistance genes, essentially by making them more potent. The other is to knock out what are known as 'susceptibility' genes. These are normal plant genes that happen to be involved in processes favouring the fungi that cause disease. By reducing their function, it is possible to develop resistance to such fungi. We are trying both approaches. Our most advanced edited Cavendish plants have been at our field trial site for more than twelve months. It's early days but some of the lines are promising. At the same time, we are also taking a completely different strategy. Goldfinger has excellent resistance to both of the major diseases but, since its fruit has not proved popular, we are editing the genes involved in fruit taste and texture with the goal of increasing consumer appeal. The first of these plants will be tested in the field by the end of 2025.

We now have multiple routes to the development of new varieties, and particularly a Cavendish replacement: conventional breeding, genetic modification and gene editing. And there are additional technologies or modifications in the pipeline. Where to from here?

What About the Rest?

One of the most fascinating banana cultures in the world is centred in Uganda. The staple food of Uganda is cooked bananas: East African Highland bananas. There are plenty of local landraces with wonderful names such Nakitembe, Nakinyika and Mpologoma; to me, they look and taste identical, but the farmers of Uganda can immediately tell them apart and describe their individual attributes. These bananas were introduced from their centres of origin way before historical records began, and the farmers have selected the different landraces over many centuries. Uganda is primarily an agrarian country; less than 20 per cent of the population of around 49 million live in the capital, Kampala. Staple foods, particularly in developing countries, are starchy foods that provide base-load energy: think maize, wheat, rice, potatoes, cassava, yams, taro, sweet potato, sorghum, and, of course, bananas when they are harvested green, before the starch is converted to sugars. In Uganda, once harvested, the fruit is peeled and steamed for a number of hours and then mashed and served possibly with a peanut or groundnut sauce and vegetables such as beans. For rural smallholders, a meal including meat is a rarity.

The average yield from East African Highland bananas grown by smallholders in Uganda is less than one third of their potential. While East African Highland is resistant to Panama Disease race 1, and TR4 hasn't reached Uganda, it's still badly damaged by Black Sigatoka, and soil fertility is a problem as the land has been farmed for centuries. The other major disease is Banana Xanthomonas Wilt (BXW),[75] which seems to have originated in bananas fairly recently, perhaps in Ethiopia. It can be quite devastating as it is easily spread from plant to plant on tools, and by moths feeding on the flowers.

So, what has been done and what is being done? Back in the 1990s, Rony Swennen of KU Leuven in Belgium led a broad and dynamic banana research programme with emphasis on Africa. His group was one of the first to develop a method to genetically modify bananas. He used a cooking banana, Bluggoe, as it was very amenable to modification. The Ugandan government, led by President Yoweri Museveni, who was part of the coalition of forces who overthrew Idi Amin, had invested in banana research and development through the National Agricultural Research Organisation (NARO) which was based at the Kawanda research station on the outskirts of Kampala. In a collaboration between NARO and KU Leuven, bananas that were modified to potentially resist Black Sigatoka were planted at Kawanda in 2007. This was one of the two earliest GM banana field trials in the world, and while the results were not positive, NARO established a reputation as a major centre for banana research and development.[76] Rony now leads the International Institute for Tropical Agriculture's conventional cooking banana breeding programme in Africa.

In East Africa, the bacterial disease BXW has proven to be a major challenge. Nearly every banana variety is susceptible to it, and when left uncontrolled it can have devastating effects. It can be controlled by leaving infested soil to lie fallow for three months or more, and by sterilising cutting tools. But, while this sounds simple, most smallholders have very limited land with no flexibility to leave any fallow.

Leena Tripathi, also of the International Institute of Tropical Agriculture but then based at Kawanda, decided to use genetic modification in an attempt to generate resistance to BXW.[77] Again, in collaboration with NARO, she transferred genes from capsicums into East African Highland bananas and, after tests in the glasshouse on small banana plants, took a number of GM lines through to field trials in Uganda. The results have been extremely positive with some lines showing excellent resistance. However, unfortunately, these bananas have not progressed to farmer release in Uganda, for reasons I shall explain later. Leena has since relocated to Kenya where she has re-established her banana biotechnology lab and is working towards the next generation of BXW-resistant varieties.

My first trip to Uganda was in 2003, when I was invited to a workshop in Kampala funded partially by the Rockefeller Foundation, who were at that time providing some funding to NARO. Up to that time most of the research collaborators in sub-Saharan Africa were based in Europe and the UK. This might have been a consequence of historical colonial connections. At that time, my research collaborations in developing countries were centred in Asia and the South Pacific and funded by the Australian Centre for International Agricultural Research, whose focus then was on our region and not Africa. At that workshop,

Debbie Delmer of Rockefeller asked if I could provide some support to help NARO develop the capacity to genetically modify East African Highland bananas, and I gladly agreed. This led to my introduction to the leader of the banana biotechnology programme, Wilberforce Tushemereirwe, and so commenced a close collaboration and friendship which has lasted more than twenty years. Tush has now formally retired, but we remain connected.

In 2004, the Bill & Melinda Gates Foundation put out an international call for expressions of interest in their new Grand Challenges in Global Health programme.[78] The concept of grand challenges can be traced back to the Grand Challenges in Mathematics proposed by David Hilbert in 1900.[79] There were fourteen original Grand Challenges in Global Health relating to a wide range of different strategies, from new vaccine concepts to new diagnostics, and from new antibiotics to control of insect vectors. The initiative was designed to stimulate development of new scientific and technological solutions to health problems in the developing world. They were designed to encourage major long-term advances rather than short-term iterative improvements.

Needless to say, there was a massive response across the fourteen challenges from the international scientific community, with probably thousands of research groups submitting proposals to meet particular ones. A much smaller number of proposals were shortlisted to submit full applications. Finally, forty-four projects were funded. The level of funding per project was in the millions of dollars and in the early stages of the initiative, all the project leaders would meet together annually and in different geographical locations. Over time, for a number of projects, funding was discontinued after the first phase.

We were funded as part of Grand Challenge No. 9 (GC#9) to 'Create a Full Range of Optimal, Bioavailable Nutrients in a Single Staple Plant Species'. The focus of our Grand Challenge was on micronutrient deficiencies, including vitamin A deficiency, iron deficiency anaemia, zinc deficiency, folate deficiency and iodine deficiency. These constitute 'hidden hunger' because very often there are no obvious visible signs of disease, at least to the untrained eye.[80] These micronutrients are essential components for many bodily functions, and deficiencies in them are usually ranked as the third most significant global public health challenge after malaria and HIV-AIDS. For instance, the number of children affected by vitamin A deficiency has been estimated at more than 300 million, mostly living in low- and middle-income countries. The figures for iron deficiency anaemia are even worse, with an estimated 30 per cent of the world's population being affected. The health impacts of these deficiencies are quite devastating.

Why do these deficiencies occur? For children aged between six months and five years, it is nearly always a result of an inadequate diet. Good sources of both vitamin A and iron are meat, particularly red meat. Plants don't contain vitamin A but do contain the precursors, alpha- and beta-carotene, which are known collectively as pro-vitamin A; it is these compounds that, for instance, give carrots their orange colour. These carotenoids are converted to vitamin A in the human liver. Iron occurs naturally in plants but usually at relatively low concentrations. In low-income agrarian countries, much of the populations depend heavily on a single staple crop, always a starch crop. These are inherently low in both iron and pro-vitamin A, and such populations usually have little meat in their diet.

There have been, and continue to be, global programmes to address these micronutrient deficiencies.

In low-income countries, donor programmes have focused on providing supplements and fortifying basic foods. Supplements are usually supplied through health clinics and are very effective. Food fortification is a little more complex. Most of us are aware of iodised salt: adding iodine to table salt helps to overcome iodine deficiencies which cause thyroid problems, as well as being a leading cause of intellectual and development disabilities (a word of caution: popular 'rock' salts may not be iodised). One very effective fortification is adding vitamin A to cooking oil. However, despite these successful strategies, deficiencies persist, particularly for 'the poorest of the poor', who don't have access to health clinics and do not buy or have access to processed, fortified food. There is also 'donor fatigue', with loss of enthusiasm for programmes that have no end.

While there have been attempts to change the crops that subsistence farmers grow, it has proved very difficult to do so. These staple crops are very much part of the culture while they are adapted to the regions where they grow and the farmers have a deep understanding of their cultivation. So the concept of biofortification was developed, which involves developing varieties of the staple crops that such smallholders already grow and eat to produce elevated levels of the required micronutrient. This can be achieved by selection, by conventional breeding and by genetic modification. A very good example is sweet potatoes. Naturally occurring, orange-fleshed sweet potatoes have very high levels of pro-vitamin A, whereas white-fleshed sweet potatoes have very little. Purple-fleshed sweet potatoes are high in anthocyanins, which also have very positive health benefits. A group called HarvestPlus led a programme to develop high-iron beans, which have been released in East Africa.[81] And then there is Golden Rice, which has been developed by genetic modification. Ingo Potrykus and Peter Beyer developed the first version, transferring a gene from daffodils, plus a bacterial gene, into rice that dramatically increased the level of pro-vitamin A, giving the rice a golden colour.[82] Eventually, the daffodil gene was replaced with a maize gene and after twenty years, Golden Rice is now growing in the Philippines and in Bangladesh. Unfortunately, Greenpeace and others are attempting to stop farmers growing Golden Rice in the Philippines because of their continued opposition to genetic modification, even for humanitarian purposes.

Back to GC#9. In 2004 we, together with NARO in Uganda and PlantVax in the US, submitted an expression of interest and finally a full application to biofortify Uganda's East African Highland bananas to alleviate vitamin A deficiency and iron deficiency anaemia. Both deficiencies occurred at unacceptable levels in rural Uganda. (The other three GC#9 projects targeted rice, cassava and sorghum.)

Our proposal was funded and, working closely with Tush and his team, we devised a plan to develop the technology in Australia using Cavendish, transferring it to Uganda for implementation in the labs at the Kawanda station. This involved frequent staff exchange and students from Tush's lab doing their PhDs at our labs at QUT. Ultimately, we took the phytoene synthase gene from a Fe'i banana called Asupina. Many of the Fe'i bananas found in Micronesia and Papua New Guinea have naturally high levels of pro-vitamin A, consequently making the flesh of the fruit a brilliant orange. We inserted the gene into the Cavendish genome in Australia, and when those plants were allowed to fruit in the field, some of the fruit had bright orange flesh. We had created a 'golden banana'.[83] We still have those bananas at QUT and we have been asked why we haven't thought of releasing them in Australia. The answer is bioavailability, or the amount of pro-vitamin A absorbed by the human body after ingesting the food. For instance, if you eat a raw carrot, only about 11 per cent of the pro-vitamin A is bioavailable, but this escalates to more than 75 per cent if those same carrots are stir fried. It is the same with bananas: the pro-vitamin A in raw bananas has very low bioavailability which again dramatically escalates with cooking.[84] So, our pro-vitamin A Cavendish are very appealing to look at but, unless cooked, which is very rare in the western world, they have no additional health benefit. Not so for East African Highland bananas; they are cooked.

We took the Asupina gene and technology we used in Cavendish in our field trials in Australia to NARO in Uganda. There, Tush's team transferred them into an East African Highland banana clone called Nakitembe, as well as a conventionally bred, disease-resistant cooking banana called M9. When we took the genetically modified Nakitembe and M9 into the field, we found lines with golden-fleshed fruit and levels of pro-vitamin A well above our target. These have been taken through multi-year field trials in four different locations in Uganda, and are now ready for release to smallholder farmers. We have selected the four best lines and these are our 'golden bananas'.[85]

Tush has already devised a very low-cost approach to disseminating the plants to resource poor farmers.

First, identify the early adopters in the villages. Second, supply these farmers with high pro-vitamin A banana plants, with the proviso that for each plant they receive, they commit to passing on a certain number of suckers from that plant to others, probably neighbours, who also commit to passing on suckers. It might be called a reverse Ponzi scheme. The scheme has worked when distributing other crop varieties that are not GM. Unfortunately, we may never know how successful the scheme might be for our bananas!

Our 'golden bananas' are genetically modified, and therein lies a problem for us in Uganda. For our bananas to be released to Ugandan farmers to grow and eat, they must be deregulated: that is, the government of Uganda must approve them as safe to grow and eat. This requires legislation, commonly referred to as a biosafety bill. When we started the project with NARO in Uganda way back in 2005, there were already advanced discussions about the development and passing of such a bill. Twenty years later, there is still no bill. It has come close a couple of times only to fail at the last hurdle. We really don't know whether our bananas can ever be passed to farmers and their families in Uganda and whether Tush's plan for distributing them would be successful, becoming a platform for other crops.

There have been other GM projects in Uganda developing disease-resistant cassava and potatoes, water-efficient maize and also insect-resistant cotton, all with significant international funding. They, too, are in limbo. Meanwhile, neighbouring countries are moving forward: after a slow start, Kenya has granted GM crop approval, and Rwanda has passed its biosafety bill. It may be possible in the future to progress our 'golden bananas' through Kenya and perhaps Rwanda. where cooking bananas are also a staple. This would be another path for Leena Tripathi's disease-resistant bananas.

While Cavendish might be the banana we are used to, this process goes to show why we mustn't forget the importance of the many other varieties in other countries. After all, bananas are a crop of the tropics, primarily grown by and feeding local populations. It's not just the Cavendish that needs to be protected.

The Future of Bananas

Although farmers in the wet tropics and sub-tropics have been growing bananas for around seven thousand years at least, it has been less than two hundred years since bananas became a world food commodity. This dramatic expansion has been based on the development of large monocultures of one variety: Gros Michel, followed by Cavendish, that was exported to North America, the UK, Europe, Russia and Japan, amongst others. In that time, we have seen the emergence, spread and ultimate devastation caused by three diseases: Panama Disease race 1, TR4 and Black Sigatoka. These diseases were almost certainly around before the1800s but probably in relatively limited geographical areas. And, in those areas, the damage was likely minimal, or was at least tolerated as part of a diverse mixture of food crops grown for the family and local market. Now the diseases are rife globally.

It has become very obvious that we will need to replace the current supermarket Cavendish sooner rather than later. So where might a Cavendish replacement come from? Conventional breeding gives us one comfortable option, with the advantage of having no technology-specific regulations governing the development and release of new varieties. The conventionally bred Cavendish replacement will take ten years to be ready for export in volume. It will be different to today's Cavendish, but how different we don't know. It will likely be resistant to both Panama Disease and Black Sigatoka. There may be three or four Cavendish replacement varieties coming from different breeding programmes. As with Cavendish, these will be 'dead-end' varieties as they will be sterile and therefore unsuitable for use as parents for future varieties.

What we can confidently predict is that in the future, there will be new diseases and other new challenges for bananas. The fungus that will cause 'Panama Disease tropical race 5' is likely out there somewhere already, waiting for humans to start moving it around the world. And the fungus causing Black Sigatoka will continue to evolve and probably overcome the resistances in the new Cavendish replacements in time. When this starts having an impact, conventional breeders will have to go back to the drawing board and invest another ten years or so of research into the next replacement, assuming resistance genes for these new disease variants can be found.

Even worse, the 'COVID-19 of bananas' could arrive. This hypothetical disease crosses the species barrier; its natural host is a wild plant or maybe a domesticated crop totally unrelated to bananas. It may already be infecting bananas somewhere in Asia, and there is unlikely to be any resistance in bananas globally.

Will we then start to consider, and accept, the more controversial technologies that will provide our supermarkets, year-round, with inexpensive fruit produced with minimal or no pesticides? These are the technologies that will allow us to move only genes rather than whole genomes from one banana to another, or from other plants to bananas, or maybe even insert synthetic genes, all without disturbing the essence of the original variety. I think we will. In fact I think we must, though synthetic genes may be too far a leap to consider in the foreseeable future.

Let's consider where these genes will come from. In the world of genomics and biotechnology, artificial intelligence has been having an impact for quite some time. It's part of what we call bioinformatics. Since the first sequence of the human genome was published in 2003, improvement in the technology for sequencing and deciphering genomes has exploded. We have even sequenced the genomes of some extinct organisms. This work represents massive amounts of data, and of course the analysis requires massive data storage and massive computing power. Using supercomputers, we are identifying genes and their functions and interactions across different organisms at an amazingly rapid rate. The outcome is that ever-increasing numbers of genes are becoming available to exploit, and this applies also to bananas. Over the next decade, I am certain we will identify new resistance genes in bananas for all the current major diseases, as well as genes for many other valuable purposes, including climate resilience.

Artificial intelligence is already taking us further in biology. Proteins are the workhorses in our cells, and it is the structure of these proteins that determines what they do and how they do it. Working out the structure of a specific protein used to be a hugely complex task. This is changing as a computer program called AlphaFold is able to predict with remarkable accuracy the structure of any protein.[86] Wonderful in itself, but let's go a little further. If we know the structure of a protein, we will be able to predict how it does its job and how its functioning might be improved. And then we will be able to alter the gene to make a protein that is slightly different, maybe more efficient, or with slightly changed function.

So, ultimately, we will have a catalogue of genes to mobilise for banana improvement: genes from other bananas or from other plants that have enhanced function, such as better resistance. It could be a gene that protects bananas against frost or extreme temperatures, or a gene which can withstand flooding or destructive storm damage. However, most of this catalogue will just gather dust if we restrict ourselves to conventional breeding.

My lab has already generated a Panama Disease TR4-resistant Cavendish by transferring a single resistance gene from a wild banana into a susceptible Cavendish, and the result has been approved for production and consumption in Australia. GM bananas are a reality.

Despite all the scientific evidence to the contrary, GM crops are often regarded with some concern, sometimes with militant rejection. Global trade in agricultural commodities is bedevilled by national regulations; everything from levels of pesticide residues, to freedom from pests and diseases, to definitions of commodities and compositional analysis . . . and of course, GM. Most of these are genuine national restrictions to ensure food, crop and environmental safety; some, however, could be interpreted as trade barriers and some are unrealistically extreme or complex. Since the mid-1990s, because the genetic modification of plants was very different to any other previous breeding technique, extensive assessments of safety for human health and the environment were implemented, first in the USA, and subsequently in many other countries. Over the ensuing thirty years, various rules governing the growing and consumption of GM crops have been implemented globally, and virtually all of them are different to one another, some dramatically so. They vary from regulating the product rather than the process in Canada, to virtually banning GM crops in New Zealand (though this is being reconsidered). And rules have to be followed both in the producing country and in the importing country.

One of the criticisms of GM crops has been the control held by large multinational, agricultural biotechnology companies. A very obvious reason for this concentration has been the costs of multi-country deregulation: only these large companies could afford such upfront investment. The upshot is that, for a crop such as bananas, the costs and timescales of first understanding and then preparing a multitude of different deregulation applications in multiple countries means that certain GM crops, for example our QCAV-4 Cavendish, will probably never be seen in a supermarket in Europe. QCAV-4 was developed by a university, not a large multinational company. If Panama Disease TR4 gets bad enough, which ultimately it will, and there are no other Cavendish alternatives, it is possible that QCAV-4 may find its way into North America and perhaps Japan. So, even with all the promising genes that will become available, it is unlikely that a strictly GM banana will gain global acceptance in the international export market in the foreseeable future. The possibility of harmonising GM regulations, particularly at a detailed level, between all or even a few countries is very remote.

There is a spark of hope: cisgenics.[87] Cisgenics involves moving genes within a plant species, in our case from one banana to another. To qualify as a true cisgenic, only the DNA as it occurs in one banana is moved to another banana. This happens in conventional breeding, the difference being that by cisgenics we can move single, desirable genes rather than the tens of thousands of genes, most of which are superfluous at best, moved by conventional breeding. There are now very positive moves in North America, Australia and even in Europe to avoid considering cisgenic plants as GM. This is a game changer. For our QCAV-4, we moved the RGA2 gene from a wild banana into Cavendish but we also moved other bits of DNA that were not from the wild banana, so QCAV-4 is not a cisgenic. However, we are in the process of making a cisgenic Cavendish which has the RGA2 gene just as it occurs in the wild banana. It should be available in five or six years.

With newly recognised resistance genes, we could easily see cisgenic Cavendish in multiple markets within a decade. The advantages over a new conventionally bred Cavendish replacement is that the cisgenic will be real Cavendish, not a replacement, and additional genes can be added to these new cisgenics at any time in the future as necessary. No need to go back to the beginning. The one technical limitation at this time is that we can't predetermine where the new cisgenes will be inserted in the genome. However, this issue will be overcome.

The other advanced genetic technology is gene editing, which is also classified or likely to be classified as non-GM in many countries, depending on how it is implemented. Therefore, in those geographies, it would have the same regulatory advantages as conventional breeding and cisgenics. It can be considered GM 'light' as the genome is modified but there is no added DNA. The other advantage is that it's highly specific. The current applications in plants are quite limited and most involve shutting down a gene, either totally or partially. To a certain extent, these applications are a 'workaround' to avoid or overcome the impost of GM regulation and, consequently, gain consumer acceptance. In animal systems, it is possible to use CRISPR editing to replace entire genes. This can't be done in crop plants efficiently as yet, but it will come.

This will be a major step forward. We will be able to combine cisgenics with CRISPR so as to insert the entire gene of our choice at the location of our choice. I think we will be doing this within the next decade and it will provide very many opportunities. Obviously, these advances won't be banana specific; they will probably impact virtually all agricultural crops. Gene-edited crops that have already been released include rice,[88] tomatoes,[89] soybeans[90] and mushrooms.[91] These are the early front runners; there are many more in development.

There is another dynamic that will influence the future of bananas. The rise of research activity from China's investment in plant sciences (and most other scientific fields) has been unfolding over the past decade, and we are now seeing the weight of publications in plant biotechnology shift from North America and Europe to China. Banana research is part of that. In the not-too-distant future, we will see that research effort leading into the development, and ultimately the production, of dynamically new banana varieties. India too is ramping up its research and development in bananas, so we are now seeing the two largest producers, both of which are part of the centres of origin, becoming major players in the future of bananas. Unfortunately I suspect that this effort will be focused on a Cavendish replacement, at least initially, rather than expanding the array of varieties.

There is nearly a clear path to the development of the next supermarket bananas. The technologies are available and constantly improving. Since Europe in particular is seriously considering accepting gene-edited and cisgenic crops as non-GM, no repressive regulation and no labelling will be needed. North America, amongst others, is essentially there. If the European proposal is accepted and ratified, then there will be an explosion of investment in developing new varieties of most crops and, in our case, bananas.

What about those varieties that make up the other approximate 50 per cent of global production, including the cooking types such as East African Highland bananas, the array of different plantains, as well as the lesser-known dessert bananas such as Prata Ana, Lady finger, Sukali Ndizi and, my favourite by name, Goly Goly Pot Pot? There are conventional breeding programmes targeting the improvement of some of these, both in Africa and South America. These new varieties, which are aimed at being disease resistant, will have the same limitations as the new, conventionally bred Cavendish replacements: they will be 'dead end' varieties. However, all the new genes and traits in our 'catalogue' will be equally available to all the non-Cavendish varieties. The challenge to develop new non-Cavendish varieties using advanced genetic technologies will not be the technologies themselves, but rather a combination of the capital investment needed and the local regulations to allow gene-edited and cisgenic bananas to be grown and eaten. The regulations must come first. Our experience in Uganda clearly demonstrates that if such a regulatory system is not in place in a particular country, then investment in advanced genetic technologies will either not materialise or will evaporate. Assuming the regulatory environment is accommodating, then our gene catalogue will be available, but funding its utilisation will be difficult. However, it will happen. Innovators, entrepreneurs and maybe local governments will take the opportunity to develop this next layer of diverse cooking and dessert bananas, not only for disease resistance but with traits such as slower ripening that would allow them to be transported to distant markets – our markets. Relatively small tweaks in the genomes of varieties that are currently confined to local markets could see them become 'exportable', not as mass commodities but as boutique varieties produced by smallholders and commanding a premium price. Imagine a 'roadside stall' in your supermarket replacing the current single offering. New opportunities for both the small developing country producers and the western consumer: a win-win.

There is one huge group of banana varieties and landraces that will probably never be 'improved' but are part of rich, local cultures, primarily in Asia.

I have never been to the highlands of Papua New Guinea but my colleagues have described many different varieties, some that 'taste like strawberries' or have other wonderful exotic flavours. The preservation of such varieties in their natural environment is in the hands of the local small farmers and communities who value them as part of their culture. We are dependent on those who have no access to advanced technology to safeguard our global biodiversity, and the less we interfere, the greater the likelihood of success.

While diseases have dominated the drive towards new varieties, future opportunities are not limited to resistance. Although, for instance, the cooking bananas we have developed with elevated pro-vitamin A in order to overcome vitamin A deficiency in East Africa are stuck in limbo for now, gene editing will likely see golden bananas in the market one day. In supermarkets today, you will find the standard white cauliflowers but also orange cauliflowers with high pro-vitamin A, and purple cauliflowers with high anthocyanins, potent antioxidants with known health benefits. These are all natural variants. In the US, you can buy Pinkglow pineapples. These are GM and have high levels of lycopene, the red colour in tomatoes, also known as an antioxidant. All these fruit 'colours' can be developed in bananas, whether for consumer appeal or for real health benefits. We could probably have a purple-fleshed banana full of anthocyanins for a fraction of the cost of blueberries, which are often promoted for their anthocyanin content.

Now, the 'elephant in the room': climate change. Unless the vast majority of scientists are wrong, which is extremely unlikely, then the next seventy-five years are going to be much more challenging for crop production, including for bananas. With rising temperatures and increasingly erratic weather patterns leading to more frequent and severe droughts, more devastating storms and flooding, our food systems will become less reliable and less productive. There are already indications of this for banana production in the Americas. The two dominant types of Cavendish, Grand Nain and Williams, differ in their performance in different climatic conditions. Grand Nain outperforms Williams when growing conditions are near ideal. The opposite is true when conditions are less than ideal. Grand Nain has been the choice in the Americas until recently. There is now a significant increase in demand from the large growers to replant with Williams. These growers have clearly recognised that conditions are changing and are making large investments, a portent of things to come. In a very recent study, a multi-disciplinary team from the UK and Costa Rica have forecast a 60 per cent reduction in the area suitable for export banana production across Latin America and the Caribbean due to rising temperatures, labour shortages and infrastructure.[92] Those developing new varieties, including my team, are only just now coming to grips with how we can build climate resilience into the next generation. Many questions remain. What traits do we 'breed' for: drought tolerance? heat tolerance? or maybe shorter stature and stronger stems to withstand damaging storms? What are the resilience genes and where will they come from?

What, then, does the future of bananas look like? The future does not depend on technologies. These are effectively a given and they are constantly being improved, invented and developed. We must be able to genetically improve bananas continuously so that they can be produced in ever more challenging environments with the minimum amount of chemical inputs, and the maximum genetic diversity. The technologies to achieve these outcomes are either already available or will be within a decade. The problem is regulation.

North America is setting the pace for crop innovation and much of South America is close behind. If Europe decides that gene-edited and cisgenic crop varieties are not to be regulated, then much of Africa will likely follow. Asia is more difficult to predict as the huge domestic food markets of India and China will independently decide whether to regulate or not. India has already released gene-edited rice. If Europe fails to adopt their proposed 'no regulation' position in this decade, or if the proposal is overly modified, at best there will be segregation, with a diversity of new banana varieties going, for instance, to North America, and a single variety going to Europe.

The future impacts of diseases and climate change must surely outweigh apprehension or indecision about harnessing advanced genetic technologies, let alone the opportunity to enjoy the amazing diversity of flavours, colours and even health benefits that are technically within our grasp. For once, let us conserve, improve and enjoy these ancient fruits of the forest rather than casting them aside for a few replacements, which themselves will eventually need to be replaced.

The choice is ours.

Acknowledgements

During my professional career, I have predominantly written for a scientific audience with fairly dry, unemotional delivery of the outcomes and implications of our research projects. These published papers are full of verifiable facts, assumed knowledge and jargon. The opportunity to address a much broader audience was both exciting and daunting. That transition of writing style did not come easily to me and I particularly thank my editor, Nikki Griffiths, who gently and patiently though consistently led me over many months to arrive at this manuscript. And my thanks equally to my wife, Ged, who read so many drafts and made so many invaluable comments and suggestions while, at the same time, being constant in her support and enthusiasm.

Our efforts to develop improved banana varieties have required the skills and expertise of many researchers over nearly forty years. They have included honours students, PhD students, postdoctoral fellows and research scientists, too numerous to name individually. However, there are two standouts. Rob (Prof. Robert Harding) joined me as a postdoctoral fellow in 1988 and evolved as my co-leader. Together, with an amazing level of agreement, we developed our targets and strategies and, against the odds, we have realised nearly all those targets. And Tush (Dr Wilberforce Tushemereirwe) was my co-leader for our 'golden banana' project in Uganda. His depth of knowledge of bananas, his commitment to food security for his country and his ability to embrace technology made him such a valuable long-term collaborator and friend.

We have been financially supported by many government and non-government organisations in our quest to develop enhanced banana varieties, but there have been a few standouts for both their support and encouragement. My university, Queensland University of Technology, has been a constant and enthusiastic backer of our research for more than thirty years. In Australia, when few would even consider being involved in GM research, banana grower Robert Borsato and ultimately marketer and grower Premier Fresh Pty Ltd looked to the future and took the risk, financing our field trials in the Northern Territory over more than a decade. And finally, the Bill & Melinda Gates Foundation had faith in our ability to deliver and funded our African programmes for nearly eighteen years. Without such support, innovation cannot flourish.

I want to acknowledge the undocumented and unheralded smallholder and subsistence farmers of Asia, the Pacific including Papua New Guinea, and Africa. It is they who are the 'holders' of the wonderful diversity of banana varieties and landraces as well as many other crops. Without their daily efforts, much of that diversity would be lost.

References

The Ancient Fruits of the Forest

1. Pennisi, E., 'Researchers have gone bananas over this fruit's complex ancestry', *Science* (14 Oct. 2022), doi: 10.1126/science.adf3420

2. 'Musa acuminata', ProMusa, https://www.promusa.org/musa+acuminata

3. Denham, T., Haberle, S., Lentfer, C., Fullagar, R., Field, J., Therin, M., Porch, N. and Winsborough, B., 'Origins of Agriculture at Kuk Swamp in the highlands of New Guinea', *Science*, 301 (11 Jul. 2003), 189–93, doi: 10.1126//science.1085255

4. Mindzie, C., Doutrelepont, H., Vrydaghs, L., Swennen, R., Beeckman, H., de Langhe, E. and de Maret, P., 'First archaeological evidence of banana cultivation in central Africa during the third millennium before present', *Vegetation History and Archaebotany*, 10 (2001), 1–6

5. 'Journey to a better Mchare', IITA, (6 Oct. 2018), https://www.iita.org/news-item/journey-to-a-better-mchare/

6. Asasira, M., Khakasa, E., Akankwasa, K., Marimo, P., Tumuhimbise, R., Mpirirwe, I., Kleih, U., Forsythe, L., Fliedel, G., Dufour, D. and Nowakunda, K., 'The East African highland cooking bananas "Matooke" preferences of farmers and traders: Implications for variety development', *International Journal of Food Science and Technology*, 56 (2021), 1124–1134
7. Langdon, R., 'The banana as a key to early American and Polynesian history', *The Journal of Pacific History,* 28 (1993), 15–35
8. 'The Origin of the Canary Island Banana is in Sitio Litre', Historical Orchid Garden, Sitio Litre, https://jardindeorquideas.com/en/the-origin-of-the-canary-island-banana-is-on-the-litre-site/

The Global Banana

9. Bell, B., 'London's 391 years of going bananas for bananas', BBC News (10 Apr. 2024), https://www.bbc.co.uk/news/uk-england-london-68770149
10. 'The Evolution of the Common Banana from Gros Michel to Cavendish', United Fruit Company, https://www.unitedfruitcompany.com/articles/the-evolution-of-the-common-banana-from-the-gros-michel-to-the-cavendish
11. Josling, T. E. and Taylor, T. G., *Banana Wars: The Anatomy of a Trade Dispute* (CABI Publishing, 2003), https://www.cabidigitallibrary.org/doi/pdf/10.5555/20033051451
12. Abbott, R., 'A socio-economic history of the international banana trade, 1870–1930', *EUI Working Paper RSCAS 2009/22* (2009)
13. 'Sam Zemurray', Wikipedia (last modified 21 May 2025), https://en.wikipedia.org/wiki/Sam_Zemurray

14. 'Banana massacre', Britannica, https://www.britannica.com/event/Banana-Massacre
15. Anderson, C., 'Florida jury finds Chiquita Brands liable for Colombia deaths, must pay $38.3M to family members', AP News (11 Jun. 2024), https://apnews.com/article/chiquita-colombia-death-squads-jury-verdict-b2519c1bc2d7eb6bea207f543d8301fd
16. Dodds, T., 'The story of the Cavendish Banana', Lightwater (1 Jan. 2023), https://lightwater.wordpress.com/2023/01/01/the-story-of-the-cavendish-banana/
17. Paxton, J., 'Musa cavendishii', *Paxton's Magazine of Botany and Register of Flowering Plants Volume 3* (1837), 51–52
18. Thompson, S., 'The quest to save the banana from extinction', The Conversation (18 Apr. 2018), https://theconversation.com/the-quest-to-save-the-banana-from-extinction-112256
19. Ryley, M. and Drenth, A., 'Joseph Bancroft's discovery of Fusarium wilt of bananas', *Historical Records of Australian Science*, 35 (6 Mar. 2024), 158–169, doi:10.1071/HR23012
20. McKenny, R. E., 'The central American banana blight', *Science*, 31 (1910), 750–751
21. Smith, E., 'A Cuban banana disease', *Science*, 31 (1910), 754–755
22. Carefoot, G. and Sprout, E., 'The great potato famine and after', in *In Famine on the Wind* (Angus and Robertson, 1969), 64–84
23. Garcia-Bastidas, F., 'Fusarium oxysporum f.sp. cubense tropical race 4 (Foc TR4)', CABI Digital Library (6 Jun. 2022), doi: 10.1079/cabicompendium.59074053

24. Marquardt, S.,'"Green Havoc": Panama Disease, Environmental Change, and Labor Process in the Central American Banana Industry', *The American Historical Review*, 106 (Feb. 2001), 49–80, https://doi.org/10.2307/2652224
25. Simmonds, N. W., 'Panama Disease, Banana Wilt', in *Bananas* (Longman, 1959), 366–378
26. Hippolyte, I., Jenny, C., Gardes, L., Bakry, F., Rivallan, R., Pomies, V., Cubry, P., Tomekpe, K., Risterucci, A., Roux, N., Rouard, Arnaud, E., Kolesnikova-Allen, M and Perrier, X.,'Foundation characteristics of edible *Musa* triploids revealed from allelic distribution of SSR markers', *Annals of Botany,* 109 (5 Apr. 2012), 937–951, doi: 10.1093/aob/mcs010
27. Jones, S., 'Spanish police seize record cocaine haul in banana shipment from Ecuador', *The Guardian* (7 Nov 2024), https://www.theguardian.com/world/2024/nov/07/spanish-police-seize-record-cocaine-haul-in-banana-shipment-from-equador

Crunch Time

28. Ploetz, R., 'Black Sigatoka of banana: The most important disease of a most important fruit', *The Plant Health Instructor*, 1 (1 Jan. 2001), doi: 10.1094/PHI-I-2001-0126-02
29. 'Black Sigatoka', Wikipedia (last accessed 4 Jun. 2025), https://en.wikipedia.org/wiki/Black_sigatoka
30. Hwang, Shin-Chuan and Ko, Wen-Hsiung., 'Cavendish banana cultivars resistant to Fusarium wilt acquired through somaclonal variation in Taiwan', *Plant Disease*, 86 (2004), 580–588, doi: 10.1094/PDIS.2004.88.6.580

31. Molina, A., Fabregar, E., Sinohin, V., Yi, G. and Viljoen A., 'Recent occurrence of *Fusarium oxysporum f.sp. cubense* tropical race 4 in Asia', *ISHS Acta Horticulturae*, 828, doi: 10.17660/ActaHortic.2009.828.10

32. 'Battling Panama disease in Philippine bananas', Global Plant Protection News (3 Jan. 2015), https://iapps2010.wordpress.com/2015/01/03/battling-panama-disease-in-philippines-bananas/

33. 'TR4 present in the UK', ProMusa (14 Dec. 2018), https://www.promusa.org/blogpost580-TR4-present-in-the-UK#:~:text=Warmington%20confirmed%20to%20ProMusa%20that,garden%20collections%2C%E2%80%9D%20says%20Warmington.

34. Martinez, G., Olivares, B., Rey, J., Rojas, J., Cardenas, J., Muentes, C. and Dawson, C., 'The advance of Fusarium wilt tropical race 4 in Musaceae of Latin America and the Caribbean: current situation', *Pathogens*, 12 (8 Feb. 2023), 277, doi: 10.3390/pathogens12020277

35. 'Banana crisis escalates after TR4 has been confirmed in Ecuador', Fresh Plaza (10 Nov. 2025), https://www.freshplaza.com/north-america/article/9783526/banana-crisis-escalates-after-tr4-has-been-confirmed-in-ecuador/

36. 'Pest Profile: Panama disease tropical race 4 in Australia', Australian Government Department of Agriculture, Fisheries and Forestry (16 Jun. 2023), https://www.agriculture.gov.au/about/news/panama-disease-tropical-race-4-in-australia

37. Harding, R., Paul, J.Y., James, A., Kleidon, J., Smith, M., Shekhawat, U., Phillips, A., Kidanemariam, D., Dawit, A. and Dale, J., 'QCAV-4, the first genetically modified Cavendish (cv. Grand Nain) banana resistant to Fusarium wilt tropical race 4 approved for commercial production

and consumption', *Plant Biotechnology Journal* (9 Jun. 2025), https://doi.org/10.1111/pbi.70178

38. 'First Dutch bananas end up as cakes, vegan meat and lingerie', Wageningen University & Research (25 Nov. 2021), https://www.wur.nl/en/newsarticle/first-dutch-bananas-end-up-as-cakes-vegan-meat-and-lingerie.htm

39. Pearce, F., 'The last days of the banana', *New Scientist* (18 Jan. 2023), https://www.newscientist.com/article/mg17723784-800-going-bananas/

40. 'Yes! We have no bananas', Wikipedia (last accessed 12 Apr. 2025) https://en.wikipedia.org/wiki/Yes!_We_Have_No_Bananas

41. Padillo, M. M., 'PHL losing share in key banana markets due to pests, diseases', BusinessWorld (30 Sep. 2024), https://www.bworldonline.com/agribusiness/2024/09/30/624562/phllosing-share-in-key-banana-markets-due-to-pests-diseases/#google_vignette

42. Programa de Banano y Plátano', FHIA, https://fhia.org.hn/banano-y-platano/

43. Vezina, A., 'Tribute to Phil Rowe', ProMusa (29 Jul. 2013), https://www.promusa.org/blogpost307

44. 'Goldfinger banana', Wikipedia (last accessed 17 Dec. 2023), https://en.wikipedia.org/wiki/Goldfinger_banana

45. 'Genomes', Banana Genome Hub, https://banana-genome-hub.southgreen.fr/

46. 'GenBank Overview', National Library of Medicine, https://www.ncbi.nlm.nih.gov/genbank/

47. 'Banana streak disease', CABI Compendium (10 Dec. 2020), https://www.cabidigitallibrary.org/doi/full/10.1079/cabicompendium.8548

48. Vezina, A., 'Banana breeder's explorer', ProMusa (4 Jul. 2015), https://www.promusa.org/blogpost299

49. D'Hont, A., et al., 'The banana (Musa acuminata) genome and the evolution of monocotyledonous plants', Nature, 488 (11 Jul. 2012), 213–217, https://www.nature.com/articles/nature11241

50. 'Banana and Plantain – What is CIRAD doing?', CIRAD, https://www.cirad.fr/en/our-activities-our-impact/tropical-value-chains/banana-and-plantain/what-cirad-is-doing

51. 'Yelloway', Chiquita, https://www.chiquita.com/yelloway/

52. 'Yelloway presents its first resistant banana plants: Yelloway One', KeyGene, https://www.keygene.com/newsitem/yelloway-banana-breeding-presents-its-first-resistant-banana-plants-yelloway-one/

53. 'Elo Life Systems announces strategic partnership with Dole Food Company with the goal of saving the Cavendish banana', Elo (4 Aug. 2020), https://elolife.com/news/saving-the-cavendish-banana/

54. 'Banana and plantain breeding program: Phase III', Embrapa (2021), https://www.embrapa.br/en/busca-de-projetos/-/projeto/218485/banana-and-plantain-breeding-program-phase-iii

55. 'ICAR: National Research Centre for Banana', India Science, Technology and Innovation Portal, https://www.indiascienceandtechnology.gov.in/organisations/ministry-and-departments/icar-national-research-centre-banana

56. 'Redesigning banana breeding', IITA, https://www.iita.org/iita-project/redesigning-banana-breeding/

The Next Generations

57. Gelvin, S. B., 'Agrobacterium-Mediated Plant Transformation: the Biology behind the 'Gene-Jockeying' Tool', *Microbiology and Molecular Biology Reviews* (1 Mar. 2003), doi: 10.1128/mmbr.67.1.16-37.2003

58. Sagi, L., Panis, B., Remy, S., Schoofs, H., De Smet, K., Swennen, R. and Cammue, B. P., 'Genetic transformation of banana and plantain (Musa spp.) via particle bombardment', *Nature Biotechnology*, 13 (13 May 1995), 481–485
59. Khanna, H., Becker, D., Kleidon, J. and Dale, J., 'Centrifugation assisted Agrobacterium tumefaciens-mediated transformation (CAAT) of embryogenic cell suspensions of banana (Musa spp Cavendish AAA and Lady finger AAB)', *Molecular Breeding*, 14 (Oct. 2004), 239–252
60. May, G., Afza, R., Mason, H., Wieko, A., Novak, F. and Arntzen, C.,'Generation of transgenic banana (Musa acuminata) plants via Agrobacterium-mediated transformation', *Nature Biotechnology,* 13 (1 May 1995), 486–492
61. Peraza-Echeverria, S, Dale, J. L., Harding, R. M., Smith, M. K. and Collet, C., 'Characterisation of disease resistance gene candidates of the nucleotide binding site (NBS) type from banana and the correlation of a transcriptional polymorphism with resistance to Fusarium oxysporum f.sp cubense race 4', *Molecular Breeding*, 22 (27 Jun. 2008), 565–579
62. Dale, J., James, A., Paul, J-Y., Khanna, H., Smith, M., Peraza-Echeverria, S., Garcia- Bastidas, F., Kema, G., Waterhouse, P., Mengersen, K and Harding, R.,'Transgenic Cavendish bananas with resistance to Fusarium wilt tropical race 4', *Nature Communications*, 18 (14 Nov. 2017), 1496, doi: 10.1038/s41467-017-01670-6
63. Harding, R., Paul, J. Y., James, A., Kleidon, J., Smith, M., Shekhawat, U., Phillips, A., Kidanemariam, D., Dawit, A. and Dale, J., 'QCAV-4, the first genetically modified

Cavendish (cv. Grand Nain) banana resistant to Fusarium wilt tropical race 4 approved for commercial production and consumption', Plant Biotechnology Journal (9 Jun. 2025), https://doi.org/10.1111/pbi.70178
64. 'Risk assessment and risk management plan for DIR199', Office of the Gene Technology Regulator (12 Feb. 2024). https://www.ogtr.gov.au/sites/default/files/2024-02/dir199_full_risk_assessment_and_risk_management_plan.pdf
65. 'A1274 – Food derived from disease-resistant banana line QCAV-4', Food Standards Australia and New Zealand (29 Apr. 2024), https://www.foodstandards.gov.au/food-standards-code/applications/A1274-Food-derived-from-disease-resistant-banana-line-QCAV-4
66. Dale, J., Harding, R., Khanna, H., James, A., Kleidon, J., Smith, M. and Shekhawat, U., 'Banana plant named QCAV-4 USPP34398P3', United States Plant Patent (2020), https://patents.google.com/patent/USPP34398P3/en
67. Zhu, H., Li, C and Gao, C. 'Applications of CRISPR-Cas in agriculture and plant biotechnology', *Nature Reviews Molecular Cell Biology*, 21 (24 Sep. 2020), 661–677
68. Sagawa, C., Assis, R and Zaini, P., 'Regulatory framework of CRISPR-edited crops in the United States', in *Global Regulatory Outlook for CRISPRized Plants* (Academic Press, 2024), 167–195
69. EFSA Panel on Genetically Modified Organisms, 'Criteria for risk assessment of plants produced by targeted mutagenesis, cisgenesis and intragenesis', EFSA Journal (20 Oct. 2022), https://efsa.onlinelibrary.wiley.com/doi/full/10.2903/j.efsa.2022.7618

70. 'CRISPR-Cas: growing more, sustainably', Corteva Agriscience, https://www.openinnovation.corteva.com/what-we-offer/crispr-cas-growing-more-sustainably.html
71. Naim, F., Dugdale, B., Kleidon, J., Brinin, A., Shand, K., Waterhouse, P. and Dale, J., 'Gene editing the phyoene desaturase alleles of Cavendish banana using CRISPR/Cas9', *Transgenic Research*, 27 (Oct. 2018), 451–460
72. Tripathi, J., Htui, V., Shah, T. and Tripathi, L., 'CRISPR/Cas9-mediated editing of DMR6 orthologue in banana (Musa spp) confers enhanced resistance to bacterial disease', *Plant Biotechnology Journal* (2 May 2021), doi: 10.1111/pbi.13614
73. Kaur, N., Alok, A., Shivani, Kumar, P., Kaur, N., Awasthi, P., Chaturvedi, S., Pandey, P., Pandey, A., Pandey, A. and Tiwari, S., 'CRISP-Cas9 directed editing of lycopene epsilon-cyclase modulates flux of b-carotene biosynthesis in banana fruit', *Metabolic Engineering* (30 Jan. 2020), doi: 10.1016/j.ymben.2020.01.008
74. 'Tropic's Non-Browning Gene-Edited Banana Cleared for Production in the Philippines Tropic Biosciences', Tropic, https://tropic.bio/tropics-non-browning-gene-edited-banana-cleared-for-production-in-the-philippines/

What About the Rest?

75. Petsakos, A., Kozicka, M., Blomme, G., Nakakawa, J., Ocimati, W. and Gotor, E., 'The potential impact of banana Xanthomonas wilt on food systems in Africa: modelling scenarios of policy and disease control', *Frontiers in Sustainable Food Systems*, 7 (14 Aug. 2023), doi: 10.3389/fsufs.2023.1207913
76. Kasozi, J. and Edwards, J., 'GM banana resistant to fungus shows promise', SciDev.Net (20 Apr. 2011), https://www.

scidev.net/global/news/gm-banana-resistant-to-fungus-shows-promise-1/
77. Tripathi, L., Mwaka, H., Tripathi, J. and Tushemereirwe, T., 'Expression of sweet pepper Hrap gene in banana enhances resistance to Xanthomonas campestris pv. Musacearum', *Molecular Plant Pathology*, 11 (30 Jun. 2010), 721–731, doi: 10.1111/j.1364-3703.2010.00639.x
78. 'Gates Foundation. Grand Challenges in Global Health – Science Magazine', Gates Foundation Global Grand Challenges, https://gcgh.grandchallenges.org/grand-challenges-global-health-science-magazine
79. Hilbert, D., 'Mathematical Problems', *Bulletin of the American Mathematical Society*, 8 (1902), 437
80. Kennedy, G., Nantel, G. and Shetty, P., 'The scourge of "hidden hunger": global dimensions of micronutrient deficiencies', Food, Nutrition and Agriculture, 32 (2003), 8–16
81. 'Iron beans', HarvestPlus, https://www.harvestplus.org/crop/iron-bean/
82. Ye, X., Al-Babili, S., Klöti, A., Zhang, J., Lucca, P., Beyer, P. and Potrykus, I., 'Engineering the provitamin A (beta-carotene) biosynthetic pathway into (carotenoid-free) rice endosperm', *Science*, 287 (January 2000), 303–305
83. Paul, J.-Y., Khanna, H., Kleidon, J., Hoang, P., Geijskes, J., Daniells, J., Zaplin, E., Rosenberg, Y., James, A., Mlalazi, B., Deo, P., Arinaitwe, G., Namanya, P., Decker, D., Tindamanyire, J., Tushemereirwe, W., Harding, R. and Dale, J. 'Golden bananas in the field: elevated fruit pro-vitamin A from the expression of a single banana transgene', *Plant Biotechnology Journal*, 15 (Apr. 2017), 520–532
84. Bresnahan, K. A., Arscott, S. A., Khanna, H., Arinatwe, G., Dale, J., Tushemereirwe, W., Mondloch, S., Tanumihardjo, J. P., De Moura, F. F. and Tanumihardjos,

S.A., 'Cooking enhances but not the degree of ripeness does not affect provitamin A carotenoid bioavailability from bananas in Mongolian gerbils', *Journal of Nutrition*, 142 (Dec. 2012), 2097–2104

85. Buah, S., Tindamanyire, J., Namanya, P., Kubiriba, J., Arinaitwe, G., Paul, J. Y., Harding, R., Dale, J. and Tushemereirwe, W., 'Field selection of elite events of East African Highland Bananas expressing elevated levels provitamin A', *Plant Biotechnology Journal* (10 Jun. 2025), doi: 10.1111/pbi.70190

The Future of Bananas

86. AlphaFold Protein Structure Database, https://alphafold.ebi.ac.uk/

87. 'Cisgenics', Wikipedia (last accessed 25 May 2025), https://en.wikipedia.org/wiki/Cisgenesis

88. Priyadarshini, S., 'India approves first genome-edited rice varieties', *Nature India* (5 May 2025), doi:10.1038/d44151-025-00078-2

89. Waltz, E., 'GABA-enriched tomato is first CRISPR-edited food to enter market', *Nature Biotechnology*, 40 (Jan. 2021), 9–11

90. 'USDA grants exemption for Amfora's gene-edited ultra-high protein soybeans', ISAAA Inc. (4 May 2024), https://www.isaaa.org/kc/cropbiotechupdate/article/default.asp?ID=20796

91. Waltz, E., 'Gene-edited CRISPR mushroom escapes US regulation', *Nature*, 532 (14 Apr. 2016), 293

92. Varma, V., Mosedale, J., Alvarez, J. and Bebber, D., 'Socio-economic factors constrain climate change adaption in a tropical export crop', *Nature Foods*, 6 (6 Mar. 2025), 343–352

About the Series

Each volume in the FUTURES Series presents a vision imagined by an accomplished writer and subject expert. The series seeks to publish a diverse range of voices, covering as wide-ranging a view as possible of our potential prospects. Inspired by the brilliant 'To-Day and To-Morrow' books from a century ago, we ask our authors to write in a spirit of pragmatic hope, and with a commitment to map out potential future landscapes, highlighting both beauties and dangers. We hope the books in the FUTURES Series will inspire readers to imagine what might lie ahead, to figure out how they might like the future to look, and, indeed, to think about how we might get there.

The FUTURES Series was originally conceived by Professor Max Saunders and Dr Lisa Gee, both of whom work at the University of Birmingham. Saunders is Interdisciplinary Professor of Modern Literature and Culture, author of *Imagined Futures: Writing, Science, and Modernity in the To-Day and To-Morrow book series*, 1923–31 (OUP 2019), and editor of *The To-day and To-morrow Reader* (Routledge, 2024), and Gee is Assistant Professor in Creative Writing and Digital Media and Research Fellow in Future Thinking.

To find out more about their Future Thinking work visit www.birmingham.ac.uk/futures

Also available in the FUTURES Series:

The Future of Fraud
by Becky Holmes

The Future of Travel
by Daniel Maurer

The Future of Gardens
by Mark Lane

The Future of the Novel
by Simon Okotie

The Future of Energy
by Richard Black

The Future of the Self
by Joanna Nadin

The Future of Songwriting
by Kristin Hersh

The Future of War Crimes Justice
by Chris Stephen

The Future of Trust
by Ros Taylor

The Future of Wales
by Rhys Thomas